STUDENT UNIT GUIDE

NEW EDITION

WJEC AS Biology Unit BY1

Basic Biochemistry and Organisation

Dan Foulder

PHILIP ALLAN

Philip Allan, an imprint of Hodder Education, an Hachette UK company, Market Place, Deddington, Oxfordshire OX15 0SE

Orders
Bookpoint Ltd, 130 Milton Park, Abingdon, Oxfordshire OX14 4SB
tel: 01235 827827
fax: 01235 400401
e-mail: education@bookpoint.co.uk
Lines are open 9.00 a.m.–5.00 p.m., Monday to Saturday, with a 24-hour message answering service.
You can also order through the Philip Allan website: www.philipallan.co.uk

© Dan Foulder 2013

ISBN 978-1-4441-8291-0

First printed 2013
Impression number 5 4 3 2 1
Year 2015 2014 2013

Cover photo: Fotolia

Typeset by Integra Software Services Pvt. Ltd., Pondicherry, India

Printed in Dubai

Hachette UK's policy is to use papers that are natural, renewable and recyclable products and made from wood grown in sustainable forests. The logging and manufacturing processes are expected to conform to the environmental regulations of the country of origin.

This material has been endorsed by WJEC and offers high quality support for the delivery of WJEC qualifications. While this material has been through a WJEC quality assurance process, all responsibility for the content remains with the publisher.

P2194

Contents

Getting the most from this book

Questions & Answers

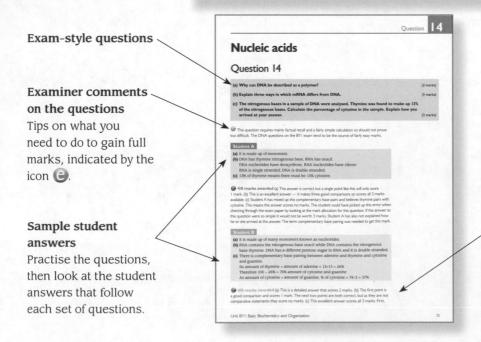

About this book

The aim of this book is to help you prepare for the WJEC AS Biology Unit BY1 examination. It will also be useful for students studying A2 biology since the concepts in BY1 are fundamental and underpin the whole of A-level biology.

The **Content Guidance** section contains everything you need to learn to cover the specification content of BY1. It can be used to reinforce your work in class as you study BY1 and also to help you prepare for the exam. There are six main topic areas in BY1 and there is a section on each of them here, as well as *examiner tips* and *knowledge checks* in the margins. Answers for these are provided towards the end of the book. At the end of each topic there is a summary of the key points covered in that topic. This is a really useful part of the book — when you are revising you can use it as a checklist and tick off each point when you have revised it. You should then not look at this topic for a few days, and *then* come back and test yourself on it to check that you really do know it.

The **Questions and Answers** section contains questions on each of the topic areas in the specification. They are written in the same style as the questions in the BY1 exam so they will give you an idea of the sort of thing you will be asked to do in the exam. After each question there are answers by two different students followed by examiner's comments on what they have written. These are important because they give you an insight into the responses the examiners are looking for in the exam. They also highlight some of the common mistakes students make.

Revision techniques

It is important to develop effective revision and study techniques. The key to effective revision is to make it *active*. Most people cannot revise effectively by just reading through notes or a book. In order to learn you have got to do something with the information. Below are some examples of active revision techniques — not all of them will work for everybody so it is important that you try them and find out which ones work for you.

Consolidate your notes

This means taking information from your notes and this book and presenting it in a different form. This can be as simple as just writing out the key points of a particular topic. One effective consolidation technique involves taking some information and turning it into a table or diagram. More creative consolidation techniques include the use of mind maps or flash cards.

The key to these techniques is that you will be actively thinking about the information you are revising. This increases the chance of you remembering it and also allows you to see links between different topic areas. Developing this kind of deep, holistic understanding of the content of BY1 is the key to getting top marks in the exam.

Complete practice exam questions

Completing practice exam questions is a crucial part of your revision. It allows you to practise applying the knowledge you have gained to the exam questions and to see if your revision is working. A useful strategy is to complete questions on a topic you have not yet fully revised. This will show you which areas of the topic you know already and which areas you need to work on. You can then revise the topic and go back and complete the questions again to check that you have successfully plugged the gaps in your knowledge.

Use technology

There are many creative ways to use technology to help you revise. For example, you can make slideshows of key points, shoot short videos or record podcasts. The advantage of doing this is that by creating the resource you are thinking about a particular topic in detail. This will help you to remember it and improve your understanding. You will also have the finished product, which you can revisit closer to the exam. You could even pass on what you have made to your friends to help them.

Content Guidance

Biochemistry

The study of biochemistry is fundamental to biology. Students often find it a difficult topic but it is important to persevere, as it can earn you some fairly straightforward marks in the exam. In this section we will first look at the biological importance of water and some key definitions. We then move on to the three main classes of chemical you need to study for the BY1 exam: carbohydrates, lipids and proteins.

Water

Water is vital to all life on Earth. Water's properties arise from the chemical structure of the water molecule. A water molecule consists of two hydrogen atoms bonded to an oxygen atom.

Water is known as a **polar molecule**. This is due to it having a slightly uneven distribution of charge. The hydrogen atoms of the water molecule are slightly positively charged while the oxygen atom is slightly negatively charged.

Knowledge check 1

Why is water described as a polar molecule?

This slight uneven distribution of charge allows **hydrogen bonds** to form between a hydrogen atom of one water molecule and the oxygen atom of another water molecule. These hydrogen bonds create a force known as cohesion that 'sticks' the water molecules together. Hydrogen bonding between water molecules is shown in Figure 1.

Knowledge check 2

Why are there cohesion forces between water molecules?

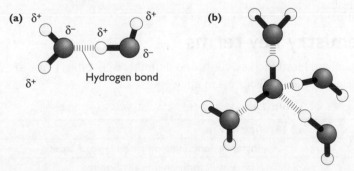

Figure 1 (a) The charges on water molecules; (b) a cluster of water molecules

Cohesion forces are responsible for the high **surface tension** of water and the formation of a 'skin' at the point where the water meets the air. This skin allows some organisms such as pond skaters and basilisks to literally walk on water. Cohesion also allows the transport of long columns of water up the xylem in the stem of plants as part of the transpiration stream.

Water is also known as a **universal solvent** as it dissolves a wide variety of different solutes. This is important for two reasons:

- It means chemical reactions can occur in solution.
- It makes transport inside living organisms much easier. Solutes are able to dissolve, for example in the blood, and then be carried around the organism.

Water is also used in a variety of chemical reactions within living organisms, for example photosynthesis.

Water has a very high **specific heat capacity**. This means that it takes a large amount of energy to raise the temperature of a body of water. The high specific heat of water is important in cells because it means that a relatively large amount of heat is required to raise a cell's temperature. This allows the cell to maintain a relatively stable internal temperature (as we will see later this is important for the actions of enzymes within the cell). The high specific heat capacity of water is also important in bodies of water such as lakes and ponds as it provides a relatively stable environment for aquatic animals.

Water has a high **latent heat of evaporation**. This means that a relatively large amount of energy is required to turn water from a liquid into a gas. This is important for living organisms as the evaporation of water (e.g. during sweating) takes energy away from the skin and causes a cooling effect.

Solid water has a lower density than liquid water. This means that solid ice floats on top of liquid water. This is important because when ice forms it provides an insulating layer on the top of a body of water. The liquid water beneath the solid ice has a higher temperature than the air above it, allowing aquatic organisms to survive even if the water at the surface has frozen.

Water's transparency is also important for living organisms. It allows light to travel through it which means that organisms such as plants and algae are able to photosynthesise.

Biochemistry key terms

It is important you learn the keywords and definitions in Table 1. You should feel comfortable using them in your exam answers.

Table 1 Biochemistry key words and definitions

Term	Definition
Element	A substance containing only one type of atom
Molecule	Two or more atoms chemically bonded
Compound	A molecule containing two or more different elements
Organic compound	A compound containing carbon and hydrogen and produced by a living organism
Inorganic compound	A compound not containing carbon and hydrogen and not produced by a living organism
Macronutrient	A nutrient that is required by a living organism in small amounts, e.g. iron and calcium
Micronutrient	A nutrient that is required by a living organism in very small quantities, e.g. copper and zinc

Knowledge check 3

Why is water known as a universal solvent?

Knowledge check 4

What property of water means that cells have a relatively consistent internal temperature?

Knowledge check 5

What is the advantage of water's high latent heat to organisms?

Examiner tip

Students often get confused between the high latent heat of water and high specific heat. Make sure you are clear about the difference between these two terms.

Examiner tip

If you are talking about the advantages to living organisms of water being transparent, make sure you are specific. Answers such as 'lets light get to plants' or 'allows fish to see' will not earn you any marks.

You also need to know some examples of macronutrients and their function within living organisms:

- iron — a component of haemoglobin which is used to carry oxygen in the blood
- calcium — used to strengthen bones and teeth
- phosphate — used to form phospholipids that make up cell membranes
- magnesium — used to form the green, photosynthetic pigment chlorophyll

Monomers and polymers

Carbohydrates and proteins form **polymers**. A polymer is a large molecule made up of repeating units called **monomers**. Polymers are formed in polymerisation reactions. These are reactions that join monomers together to make a polymer. In the polymers you need to study in BY1, **condensation** reactions join the monomers together. **Hydrolysis** reactions break the bonds in a polymer, releasing the monomers.

Carbohydrates

Carbohydrates are extremely important chemicals in biology. They range from simple sugars like glucose to large complex polymers such as chitin. All carbohydrates contain the elements:

- carbon
- hydrogen
- oxygen

It is important to know the number of bonds that each of the atoms above form in a molecule. This information will then allow you to check any structural diagrams you have drawn to make sure they are correct. Carbon atoms form four bonds, hydrogen forms one bond and oxygen atoms form two bonds.

Monosaccharides

The basic unit of a carbohydrate is a **monosaccharide**. All monosaccharides have the general formula:

$C_nH_{2n}O_n$

This means that in a monosaccharide there is the same number of carbon atoms as oxygen atoms and twice as many hydrogen atoms as carbon atoms.

Monosaccharides can be classified by the number of carbon atoms they contain:

- 3 carbon atoms — triose sugars
- 5 carbon atoms — pentose sugars (e.g. deoxyribose found in DNA)
- 6 carbon atoms — hexose sugars (e.g. glucose)

In BY1 we focus on the hexose sugars and in particular glucose. Other hexose sugars include fructose and galactose.

Glucose

We focus in detail on glucose because of its fundamental importance in biology. It is used in respiration to produce ATP (chemical energy currency used in cells) and it is the monomer for many different polysaccharides. Glucose has two isomers: alpha glucose and beta glucose. Isomers are molecules that have the same chemical

Examiner tip
The carbon atoms are not shown in the diagrams of glucose in Figure 2. This is a common, simplified way of drawing glucose. The carbon atoms are where the numbers are on the diagram. The numbers on the diagram are important because they allow you to identify which carbon atoms in a molecule form bonds.

Examiner tip
Students often have trouble explaining the difference between alpha glucose and beta glucose. Saying that the H and OH are arranged differently on carbon 1 is better than talking about up, down etc.

Examiner tip
When drawing a diagram to show a condensation reaction forming a bond, do not forget to draw in the water molecule that is produced — it is often worth a mark.

Knowledge check 7
What would be the products of the hydrolysis of maltose?

formula but a different structure. The two isomers of glucose are shown in Figure 2. Glucose can be drawn in a linear form but it is nearly always shown in the ring structure as in Figure 2.

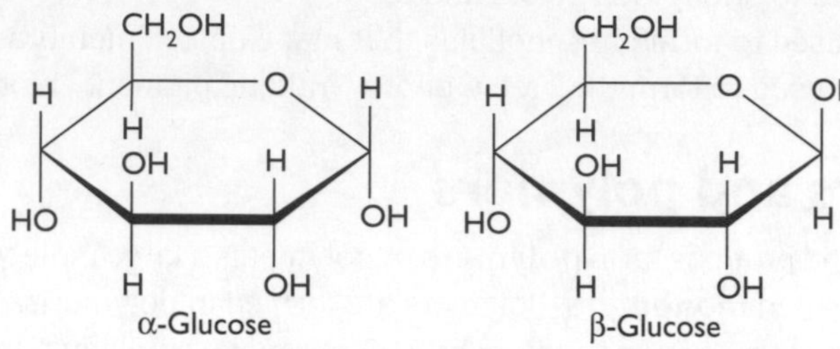

Figure 2 Isomers of glucose

The only difference between the structures of alpha and beta glucose is the different arrangement of the H and OH (hydroxyl group) on carbon 1.

Disaccharides

Two monosaccharides can be joined by a glycosidic bond to form a disaccharide. This reaction is known as a **condensation reaction**. In a condensation reaction a molecule of water is produced when the bond is formed. The bond shown is called a 1-4 glycosidic bond because it forms between carbons 1 and 4. When two α-glucose molecules are joined by an α-1-4 glycosidic bond the disaccharide **maltose** is formed. This reaction is shown in Figure 3.

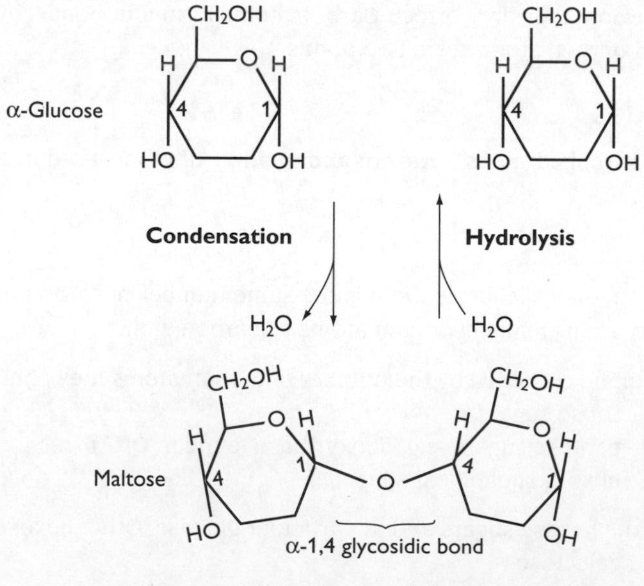

Figure 3 The formation and hydrolysis of maltose

A **hydrolysis reaction** can be used to break the chemical bond formed in a condensation reaction. In a hydrolysis reaction the bond is broken by the chemical insertion of water. If the glycosidic bond in maltose is broken by a hydrolysis reaction, two molecules of glucose will be formed.

Table 2 shows the disaccharides formed when different monosaccharides are joined by condensation reactions.

Table 2 Disaccharides formed when different monosaccharides are joined by condensation reactions

Monosaccharides	Disaccharide
Glucose + glucose	Maltose
Glucose + fructose	Sucrose
Glucose + galactose	Lactose

Polysaccharides

Three or more monosaccharides can be joined by condensation reactions to form a polysaccharide. The monosaccharides are the monomers and the polysaccharide is the polymer. Most polysaccharides contain thousands of monomers.

Polysaccharides generally either have a structural function or are used to store glucose. The large size of polysaccharides means they are insoluble; this is important for them to be able to carry out their functions. They are also osmotically inactive. This means that they can be stored in cells without any detrimental osmotic effects. In the case of the storage polysaccharides it is also important that glucose can be removed easily.

Knowledge check 8

Why is it important that polysaccharides that have a structural function are insoluble?

Storage polysaccharides

The two main storage polysaccharides are:
- **Starch** — a polymer made up of alpha glucose monomers. Starch has two components, amylose and amylopectin. Amylose is a chain of glucose monomers joined by 1-4 glycosidic bonds and formed into a helix. The structure of amylose is shown in Figure 4.
  Amylopectin has both 1-4 and 1-6 glycosidic bonds. This gives it a branched structure. The structure of amylopectin is shown in Figure 5.
  Starch is used to store glucose in *plants*.

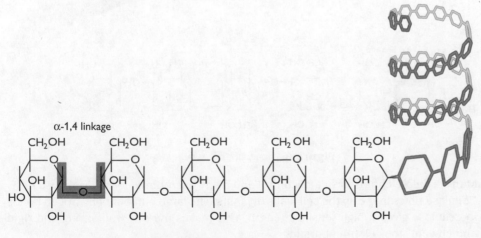

Figure 4 The structure of amylose

Content Guidance

Knowledge check 9

Which polysaccharide is used for storage in animals?

- **Glycogen** — a polymer made up of alpha glucose monomers. Like amylopectin it has both 1-4 and 1-6 glycosidic bonds so has a branched structure.
  Glycogen is used to store glucose in animals. It is found in muscle cells.

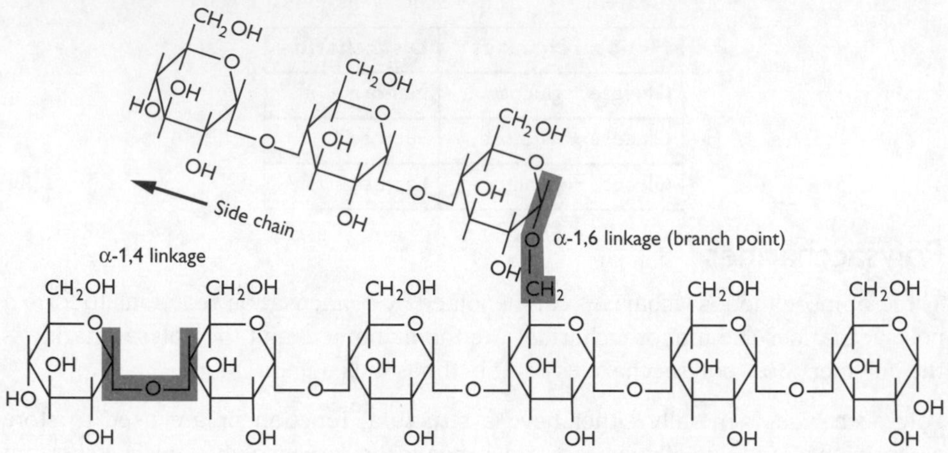

Figure 5 The structure of a branched polysaccharide: amylopectin and glycogen

Cellulose

Cellulose is a polymer made up of beta glucose monomers. Cellulose makes up the cell walls of plant cells. The beta glucose monomers are joined by 1-4 glycosidic bonds and are arranged into long straight chains. Each beta glucose molecule is rotated 180 degrees from the previous molecule in the chain. This enables hydrogen bonds to form between the OH groups in adjacent chains. The structure of cellulose is shown in Figure 6.

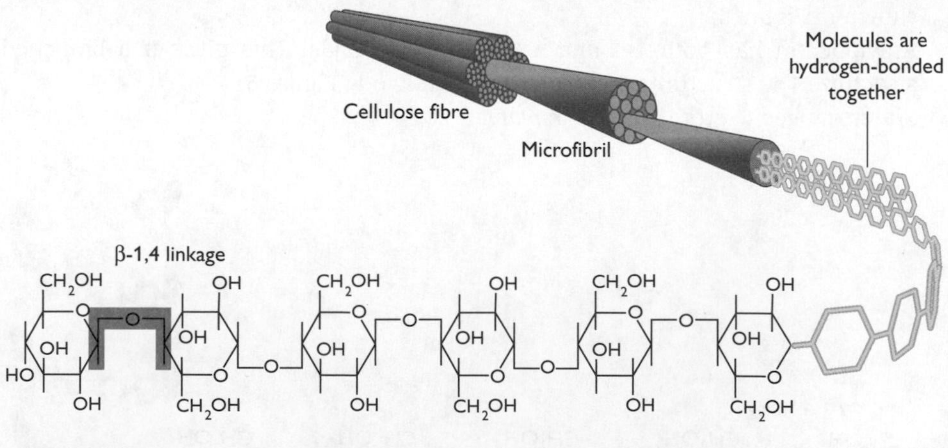

Knowledge check 10

Which monomer makes up cellulose?

Figure 6 The structure of cellulose

Many cellulose chains form a micofibril and many microfibrils form a cellulose fibre. Cellulose fibres make up the cell walls of plants. The large number of hydrogen bonds in cellulose gives it high tensile strength. This makes the cell wall strong and rigid and prevents the cell from bursting.

Knowledge check 11

What bonds give cellulose fibres their strength?

12   WJEC AS Biology

Chitin

Chitin is a mucopolysaccharide. It is made up of chains of beta glucose monomers with amino acid side chains. One OH group of each beta glucose molecule is replaced with an amino acid. Chitin is strong and lightweight and is used to form the exoskeletons of insects.

Biochemical tests for carbohydrates

You need to know the biochemical tests used to identify carbohydrates.

Starch test
- Add several drops of iodine.
- If the solution turns blue/black then starch is present.

Reducing sugar test
- Add Benedict's reagent to the unknown sample.
- Boil the solution.
- If a reducing sugar is present, a brick red precipitate will form.

All monosaccharides are reducing sugars and so are some disaccharides such as maltose.

Non-reducing sugar test

If the reducing sugar test returns a negative result, a further test can be carried out to determine whether a non-reducing sugar (such as sucrose) is present in the sample. A non-reducing sugar can be identified by first hydrolysing the glycosidic bond in the molecule, thus forming two monosaccharides. The monosaccharides will then produce a positive result when boiled with Benedict's reagent. The glycosidic bond is hydrolysed by heating the non-reducing sugar with acid.
- Heat the solution with acid.
- Neutralise by adding sodium hydroxide.
- Add Benedict's reagent to the unknown sample.
- Boil the solution.

If a non-reducing sugar was present in the original sample a brick red precipitate will form.

Lipids

Lipids are made up of the same elements as carbohydrates:
- carbon
- hydrogen
- oxygen

Unlike carbohydrates lipids are not made up of monomers that link together to form polymers. Instead lipids consist of two different molecules: glycerol and fatty acids. Glycerol is found in all lipids (Figure 7).

Examiner tip
It is easy to get confused with which monomer makes up which polysaccharide. A simple way to remember it is that the storage polysaccharides have alpha glucose monomers while the structural polysaccharides have beta glucose monomers.

Examiner tip
Students often do not bother to learn the biochemical tests but they do come up in exams. For each test you need to make sure you know:
- the name of the test
- how to carry out the test
- what the positive result of the test is

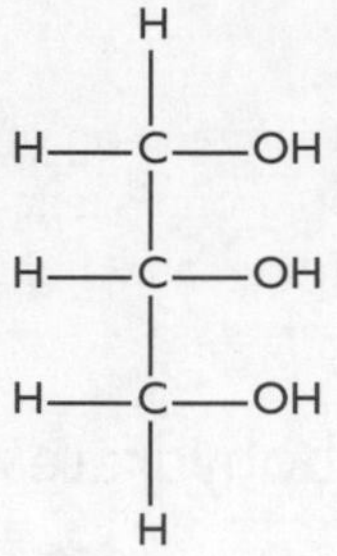

$$
\begin{array}{c}
H \\
| \\
H-C-OH \\
| \\
H-C-OH \\
| \\
H-C-OH \\
| \\
H
\end{array}
$$

Figure 7 Glycerol

Fatty acids consist of a methyl group (CH$_3$), a variable length hydrocarbon chain (CH$_2$)$_n$ and a carboxyl group (COOH). The hydrocarbon chain contains an even number of carbon atoms between 14 and 22. Fatty acids are either **saturated** or **unsaturated**. Saturated fatty acids have no carbon to carbon double bonds in the hydrocarbon chain. Unsaturated fatty acids do have carbon to carbon double bonds within the hydrocarbon chain. An example of saturated and unsaturated fatty acids is shown in Figure 8.

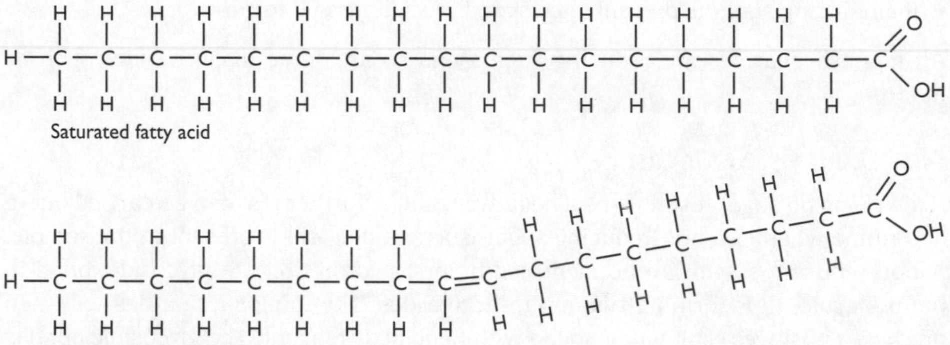

Saturated fatty acid

Unsaturated fatty acid

Figure 8 A saturated fatty acid and an unsaturated fatty acid with one double bond

Saturated fatty acids are found in animal fat and unsaturated fatty acids are found in plant oils.

Three fatty acids combine chemically with one glycerol molecule to form a **triglyceride**. There are three condensation reactions (one for each fatty acid) which form three **ester bonds**. In Figure 9 the fatty acid's hydrocarbon chain is drawn in a simplified form as a zigzagging line. In this reaction each OH group on the glycerol molecule loses a hydrogen atom while the carboxyl group of the fatty acid loses an OH to form three molecules of water. This reaction and the hydrolysis reaction that breaks the ester bonds are shown in Figure 9.

Knowledge check 13

How does a saturated fatty acid differ from an unsaturated fatty acid?

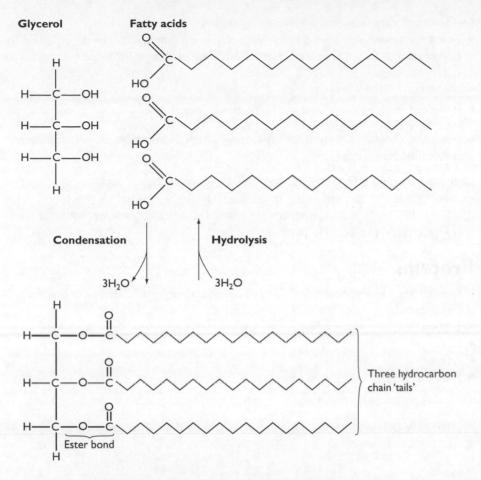

Figure 9 The formation and hydrolysis of a triglyceride

Phospholipids are a special type of lipid. Two fatty acids combine with a glycerol molecule. The other OH group on the glycerol forms a bond with a phosphate group. This makes the head (glycerol and phosphate) of the molecule hydrophilic and the tail (the two fatty acids) hydrophobic. This property is important in the formation of cell membranes. Figure 10 shows the structure of a phospholipid.

Two hydrocarbon chain 'tails'

Glycerol–phosphate 'head'

Figure 10 A phospholipid

Examiner tip

You will not be asked to draw out the whole triglyceride structure but you could be asked to show how the fatty acids bond with the glycerol, so make sure you learn this structure. It is important you also remember that the water molecules formed are made from the OH group of the fatty acid and the H from the glycerol.

Knowledge check 14

What is the bond in a triglyceride called?

Lipids have a variety of functions in living organisms:
- energy storage — a lipid stores twice as much energy as the same mass of carbohydrates; this makes it a very efficient store of energy; lipids are used in seeds to store energy
- protection of delicate organs
- thermal insulation
- buoyancy in aquatic organisms
- a source of metabolic water for organisms which live in low water environments, for example camels

Lipids are insoluble in water but soluble in organic solvents such as acetone and ethanol. Saturated fats are found in animals and are a contributory factor in heart disease. They are solid at room temperature. Unsaturated oils are liquid at room temperature and are found in plants.

Proteins

Proteins are an incredibly varied class of biological molecules. They are made up of the elements:
- carbon
- hydrogen
- oxygen
- nitrogen

They sometimes contain sulphur.

Proteins are polypeptides. A polypeptide is a polymer of **amino acids**. Figure 11 shows an amino acid.

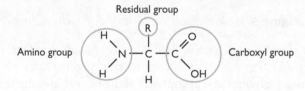

Figure 11 The general structure of an amino acid

Amino acids are made up of an amino group (NH_2), a carboxylic group (COOH) and a variable residual or R group. There are 20 different R groups. This gives 20 different amino acids. Two amino acids with different R groups are shown in Figure 12.

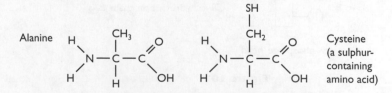

Figure 12 Alanine and cysteine

Two amino acids combine in a condensation reaction to form a **dipeptide**. A **peptide bond** forms between the carboxyl group of one amino acid and the amine group of the other amino acid. As this is a condensation reaction a molecule of water is formed. The formation of a dipeptide is shown in Figure 13.

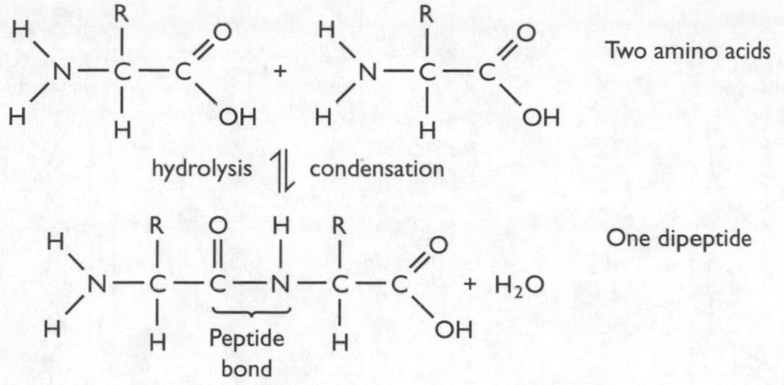

Figure 13 A condensation reaction forms a dipeptide from two amino acids

Primary, secondary, tertiary and quaternary structure of proteins

Three or more amino acids joined by peptide bonds form a **polypeptide**. The amino acids are the monomers and the polypeptide is the polymer. The type, number and sequence of amino acids in a polypeptide is its **primary structure**.

The polypeptide can twist to form either an **alpha helix** or a **beta pleated sheet**. This is the **secondary structure** of the polypeptide. The secondary structure's shape is held in place by hydrogen bonds between the peptide bonds.

The polypeptide can fold into specific complex three-dimensional shapes. This is the tertiary structure of the polypeptide. This specific structure is maintained by hydrogen bonds, ionic bonds and disulphide bonds between the R groups of the amino acids. These bonds along with a summary of primary and secondary structures are shown in Figure 14.

Knowledge check 16

What is the primary structure of a polypeptide?

Knowledge check 17

What are the two types of secondary structure?

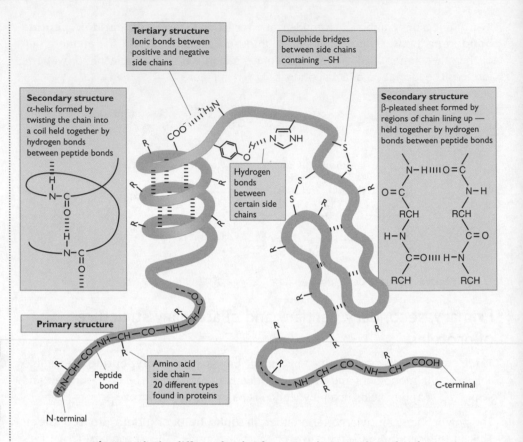

Tertiary structure
Ionic bonds between positive and negative side chains

Disulphide bridges between side chains containing –SH

Secondary structure
α-helix formed by twisting the chain into a coil held together by hydrogen bonds between peptide bonds

Secondary structure
β-pleated sheet formed by regions of chain lining up — held together by hydrogen bonds between peptide bonds

Hydrogen bonds between certain side chains

Primary structure

Amino acid side chain — 20 different types found in proteins

Peptide bond

N-terminal

C-terminal

Figure 14 The different levels of structure in a protein molecule

Two or more polypeptides with a tertiary structure can combine to form a **quaternary structure**. An example of a protein with a quaternary structure is haemoglobin. Haemoglobin is made up of four polypeptides (two alpha chains and two beta chains) and prosthetic (non-protein) iron-containing haem groups. Haemoglobin is used to transport oxygen in red blood cells. A molecule of haemoglobin is shown in Figure 15.

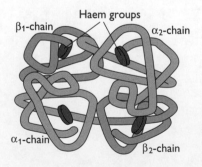

Haem groups

β_1-chain

α_2-chain

α_1-chain

β_2-chain

Figure 15 A haemoglobin molecule

Fibrous and globular proteins

Generally the proteins you study in BY1 can be divided into two types: **fibrous** or **globular**.

- **Fibrous proteins** are structural proteins that have a secondary structure. Fibrous proteins are insoluble. Examples of fibrous proteins include keratin and collagen. Keratin is used to form hair and nails and collagen is used to form tendons. Collagen consists of three alpha helix chains formed into long strands (Figure 16).

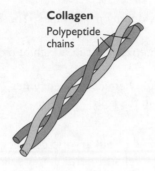

Collagen
Polypeptide chains

Figure 16 A collagen molecule

Examiner tip
Students always seem to find it difficult to remember the features of a fibrous protein — they are quite straightforward, you just need to learn them.

- **Globular proteins** have a tertiary or quaternary structure. They are soluble in water and examples include enzymes and hormones.

Biochemical test for proteins

Proteins can be identified using the biuret test. A small volume of biuret solution is added to a sample. If the sample contains protein it will turn from blue to violet.

After studying this topic you should be able to:

- Explain the biological importance of water and relate some of its properties to the fact that it is a polar molecule that forms hydrogen bonds.
- Define the key biochemistry terms and give examples of some macronutrients and their importance to living organisms.
- Explain bond formation and breaking by condensation and hydrolysis reactions.
- Describe carbohydrates in terms of monosaccharides, disaccharides and polysaccharides and explain the structure of examples of each.
- Describe the methods of the biochemical tests for carbohydrates and be able to identify the positive results.
- Explain how triglycerides form from fatty acids and glycerol and describe the differences between a saturated and unsaturated fatty acid.
- Give examples of the functions of lipids in living organisms.
- Explain how a phospholipid differs from a triglyceride.
- Describe the structure of amino acids, dipeptides and polypeptides.
- Explain the four levels of protein structure and give the bonds found in each level.
- Distinguish between fibrous and globular proteins.
- Describe the biochemical test for proteins.

Summary

Cell structure and organisation

Cells are the basic unit of a living organism. The cells you need to study can be divided into two types: **eukaryotes** (animal and plant cells) and **prokaryotes** (bacteria).

Eukaryotes

Eukaryotes are internally divided by membranes. These membranes are important because they:

- provide a surface to which enzymes can attach and on which chemical reactions can occur
- contain potentially harmful chemicals or enzymes, stopping them damaging or breaking down structures in the cell
- act as a transport system

There are many different types of eukaryotic cell. For the BY1 exam you need to study animal cells and plant cells in detail. Figure 17 shows the structure of an animal cell.

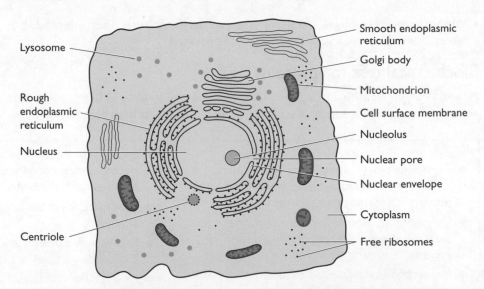

Figure 17 The ultrastructure of an animal cell

Eukaryotic cells have an exterior cell or plasma membrane that encloses the cytoplasm of the cell. The cytoplasm is a gel-like substance that contains the cell's organelles. Organelles are internal structures within the cell.

Organelles in animal and plant cells

Both plant and animal cells contain the following organelles.

Nucleus

The nucleus contains the cell's **DNA**. DNA is genetic material that is passed on from one generation of an organism to the next and provides the code for the synthesis of proteins. For the majority of a eukaryote cell's life the DNA remains inside the cell's nucleus. To allow protein synthesis to occur a strand of **messenger RNA** (mRNA) is made using DNA as a template. This mRNA strand can then leave the nucleus and carry the code for protein synthesis to the cytoplasm.

A nuclear membrane (also called the **nuclear envelope**) encloses the nucleus. This double membrane contains small 'gaps' known as nuclear pores. These nuclear pores allow the mRNA to exit the nucleus and travel to the cytoplasm. The interior of the nucleus contains a cytoplasm-like material known as the nucleoplasm. Within the nucleoplasm is the cell's DNA in the form of **chromatin**.

The nucleus also contains the **nucleolus**. The nucleolus produces rRNA (ribosomal RNA).

Mitochondria

Mitochondria (singular **mitochondrion**) produce ATP during aerobic (requiring oxygen) respiration. Like the nucleus mitochondria have a double membrane. This double membrane consists of an outer membrane, an intermembrane space and an inner membrane. The inner membrane is folded to form the cristae (singular crista). These folds increase the surface area on which ATP synthesis can occur.

Inside the mitochondria is a cytoplasm-like **matrix**. The matrix contains ribosomes and the mitochondrial DNA. This DNA allows the mitochondria to divide to meet the needs of the cell separate from the normal replication of the cell cycle. This mitochondrial DNA is evidence that the mitochondria may have once been free living organisms that were ingested by the ancestors of eukaryote cells and have lived inside eukaryotes ever since. Figure 18 shows the structure of a mitochondrion.

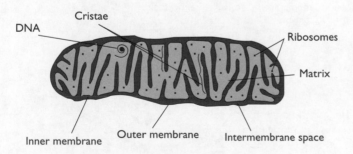

Figure 18 The structure of a mitochondrion

> **Knowledge check 19**
>
> What is the function of mitochondria?

> **Knowledge check 20**
>
> Why is it important that the cristae give the mitochondria a large internal surface area?

Rough endoplasmic reticulum

The rough endoplasmic reticulum consists of a series of membranes that are linked to the nuclear membrane. The rough endoplasmic reticulum is used as a transport system for proteins and has **ribosomes** along its length (that is why it is known as the 'rough' endoplasmic reticulum). The ribosomes carry out protein synthesis. Ribosomes are also found free in the cytoplasm.

Ribosomes

These have two subunits, large and small. The subunits come together around a strand of mRNA which fits into the mRNA groove of the ribosome. Ribosomes are made up of protein and ribosomal RNA (rRNA).

Smooth endoplasmic reticulum

The smooth endoplasmic reticulum is a series of membranes that do not have ribosomes on their surface (hence smooth). The smooth endoplasmic reticulum is involved in the synthesis of lipids.

The Golgi body

The Golgi body (also known as the **Golgi apparatus**) is a stack of flattened membranous sacs. The Golgi body has a variety of functions, including the formation of glycoproteins and lysosomes, but its main role is modifying and packaging proteins to be exported from the cell. This process is described below and shown in Figure 19.

- Vesicles (small membrane-bound sacs) containing proteins formed by the rough endoplasmic reticulum fuse at one end of the Golgi sacs.
- The protein is modified inside the Golgi sacs.
- The modified protein is then budded off in a vesicle at the other end of the Golgi sacs.
- The vesicle containing the modified protein then travels to the cell's outer plasma membrane where the protein is released by exocytosis. The process of exocytosis is discussed in more detail later (page 30).

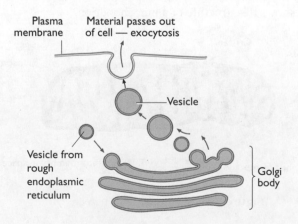

Figure 19 Golgi body

Organelles in animal cells

Animal cells also contain the following organelles.

Lysosomes

These are vesicles containing digestive enzymes. They can be used to break down worn-out organelles and to digest material taken in by phagocytosis. The action of a lysosome is explained below:

- The material is taken into the cell by endocytosis and trapped in a vacuole.
- The lysosomes fuse with the membrane of the vacuole and release their digestive enzymes into the vacuole.
- The digestive enzymes break down the material.

Centrioles

Centrioles form the spindle fibres during cell division (discussed in more detail later on page 46).

Organelles in plant cells

Plant cells contain all the above organelles except lysosomes and centrioles. They also have the additional organelles described below. The structure of a plant cell is shown in Figure 20.

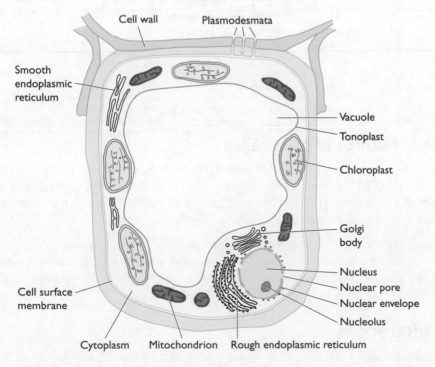

Figure 20 A generalised plant cell viewed with the electron microscope

Chloroplasts

Chloroplasts are large organelles with a double membrane (like the nucleus and mitochondria). Photosynthesis occurs in the chloroplasts. Photosynthesis is the process by which sugars and other organic molecules are formed from carbon dioxide and water using energy from sunlight.

Within the chloroplasts is a cytoplasm-like material called the stroma. Within the stroma are many membrane-bound compartments called **thylakoids**. The membrane of the thylakoids contains the chemical **chlorophyll**. Chlorophyll absorbs the light energy that is used in photosynthesis. Thylakoids form stacks called grana (singular granum). These stacks are linked by lamellae (singular lamella). The stroma also contains starch grains and ribosomes.

Like mitochondria the chloroplasts also contain ribosomes and chloroplast DNA. As in mitochondria, this chloroplast DNA is evidence that chloroplasts were once free-living organisms ingested by the ancestors of plant cells. Figure 21 shows the structure of a chloroplast.

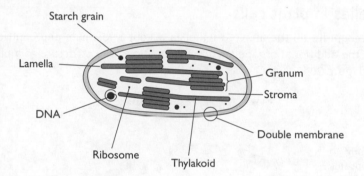

Figure 21 The structure of a chloroplast

Other features of plant cells

Plant cells also have a cellulose cell wall. This keeps the plant cell rigid and prevents it from bursting. The plant cell wall has many pores called **plasmodesmata**. The plasmodesmata allow the cytoplasm of neighbouring cells to connect, enabling substances to pass between plant cells.

Plants cells also have a large permanent **vacuole** which contains cell sap. It is surrounded by a membrane called the **tonoplast.** The vacuole is used for storage and helps to support the plant cell. Animal cells may also contain vacuoles but they will be much smaller and will not have a **tonoplast**.

Prokaryotes

Prokaryotes are cells that contain no membrane-bound organelles. Their DNA is not enclosed in the nucleus but is free in the cytoplasm. Most prokaryotes are significantly smaller than eukaryotic cells, 1–10 µm in length as opposed to 10–100 µm for eukaryotic cells. Bacteria are examples of prokaryotes. Figure 22 shows the features of a prokaryote.

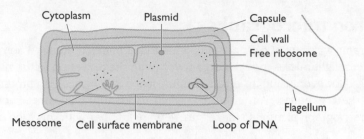

Figure 22 Features of a prokaryote

Prokaryotes have the following features:

- **capsule/slime layer** — outer layer
- **murein cell wall** — the cell wall of a prokaryote is not made from cellulose but from murein (peptidoglycan). The cell wall prevents prokaryote cells from bursting
- **mesosome** — an infolding of the prokaryotes' plasma membrane. The infolding increases the surface area for respiration and other chemical reactions to occur
- **DNA** — the DNA is free in the cytoplasm in an area known as the nucleoid. Prokaryote DNA is usually found in a single circular chromosome. Prokaryotes may also have small circular pieces of DNA called plasmids
- **ribosomes** — like eukaryotes, prokaryotes contain ribosomes. However, the ribosomes of prokaryotes are significantly smaller than the ribosomes found in eukaryotes
- **flagellum** — prokaryotes also sometimes have a flagellum, which allows them to move

Examiner tip
A common exam question is to compare the structures of eukaryotes and prokaryotes. Make sure you learn the structures of both thoroughly and do not make any obvious mistakes, for example saying they both have cellulose cell walls.

Viruses

Viruses contain no cytoplasm so are not considered cells. In fact they are generally not even thought to be alive. Viruses consist of nucleic acid and a protein coat. Viruses infect both prokaryotes (bacteriophages) and eukaryotes.

Tissues and organs

As previously mentioned the basic unit of a living organism is the cell. Cells that have a similar structure and the same function join together to form tissues. Examples of animal tissues include:

- **epithelial tissue**, e.g. cuboidal and ciliated epithelia — epithelial tissue lines spaces in animals such as the digestive and respiratory systems
- **muscle**, e.g. striated and smooth muscle — muscle tissue contracts and relaxes to move parts of animals
- **connective tissue**, e.g. collagen — structural tissue in animals

Xylem and phloem are examples of plant tissues. Tissues can combine to form organs to carry out a particular function. Examples of organs in animals include the brain and the heart. Roots and leaves are examples of plant organs. Organs form organ systems, for example the respiratory or digestive system in mammals.

Electron micrographs

The diagrams we draw of cells are simplified to aid understanding. When observing cells under a light microscope the organelles (with the exception of the nucleus) are not visible. However, using an electron microscope it is possible to produce images showing the organelles. You could be asked to identify structures shown on an electron micrograph picture so make sure you are familiar with this type of image.

Summary

After studying this topic you should be able to:
- Explain the importance of internal membranes to eukaryote cells.
- Label all the organelles found in animal and plant cells and all the parts of a prokaryote.
- Explain the functions of the different parts of the nucleus, chloroplast and mitochondria and label these parts on a diagram.
- Explain the functions of the rough endoplasmic reticulum, ribosomes, smooth endoplasmic

reticulum, Golgi body, lysosomes, centrioles, vacuole, cell wall and plasmodesmata.
- Explain the differences in structure between prokaryote cells and eukaryote cells with particular emphasis on the lack of membrane-bound organelles in the prokaryote and the fact its DNA is free in the cytoplasm.
- Describe the basic structure of a virus.
- Describe tissues and organs and give specific examples of tissues found in animals and plants.

Cell membranes and transport

The plasma membrane

The plasma membrane or cell surface membrane is the barrier through which all matter entering cells must pass. It has the following functions:
- to give the cell structure
- to allow substances to exit and enter the cell
- cell–cell recognition and cell–cell signalling

The plasma membrane is selectively permeable. This means it is able to control what enters and exits the cell.

The plasma membrane mainly consists of phospholipid molecules arranged into a bilayer (a double layer). Under the electron microscope and when stained with a water-soluble dye the cell membrane appears as three layers. This is because the hydrophilic heads of the phospholipid molecules take up the dye. The distance across the membrane is 7–8 nm.

The hydrophilic heads of the phospholipid molecules face outwards and the hydrophobic tails face inwards. This forms a hydrophobic, non-polar region in the middle of the bilayer. This prevents polar molecules such as glucose from passing through. The structure of the plasma membrane is shown in Figure 23.

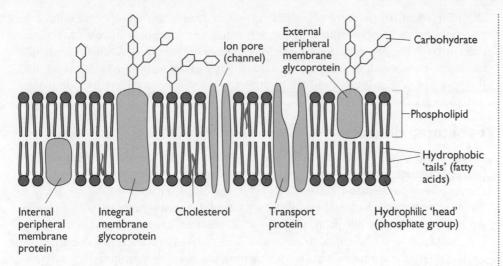

Figure 23 The structure of the plasma membrane

The plasma membrane also contains proteins. They are divided into two types:

- **intrinsic proteins** lie across both the layers of the membrane (e.g. a carrier protein)
- **extrinsic proteins** are either in one layer of the membrane or on the surface of the membrane (e.g. a glycoprotein)

Glycoproteins are proteins with a carbohydrate chain attached. Glycolipids are lipids with a carbohydrate chain attached. Both glycoproteins and glycolipids are involved in cell–cell recognition.

The membrane also contains cholesterol. Cholesterol helps to increase the rigidity of the membrane. When temperature increases, the molecules that make up the membrane gain kinetic energy and move at a faster rate. This causes the membrane to become more fluid (the components are more free to move) and therefore more permeable. In an experimental setting, organic solvents (such as ethanol) can be used to dissolve the phospholipids and so increase the permeability of the membrane.

The plasma membrane is best described by the **fluid mosaic model** proposed by Singer and Nicholson:

- fluid — all parts of the membrane can move relative to each other
- mosaic — proteins are dotted throughout the membrane like mosaic tiles

Transport across the membrane

There are several ways in which substances can move across the plasma membrane.

Diffusion

Diffusion is the movement of molecules or ions from an area of higher concentration to an area of lower concentration (down a concentration gradient).

Knowledge check 23

How does an intrinsic protein differ from an extrinsic protein?

Diffusion occurs through the phospholipid bilayer. Due to the hydrophobic tails of the phospholipids only lipid-soluble molecules, which are non-polar and uncharged, can pass through the phospholipid bilayer. Proteins are not involved. As the movement is down the concentration gradient chemical energy in the form of ATP is not needed. Examples of molecules that diffuse through the phospholipid bilayer include oxygen and carbon dioxide.

Facilitated diffusion

Large, water-soluble, polar and charged molecules or ions cannot pass through the hydrophobic tails of the phospholipid bilayer so they must move through the plasma membrane by **facilitated diffusion**. Facilitated diffusion involves the molecule or ion moving through a carrier protein or a hydrophilic pore within a channel protein. As this is a type of diffusion, movement of the molecules is again from a high concentration to a low concentration and does not require chemical energy in the form of ATP. An example of a molecule that moves across the membrane by facilitated diffusion is glucose. Diffusion and facilitated diffusion are shown in Figure 24.

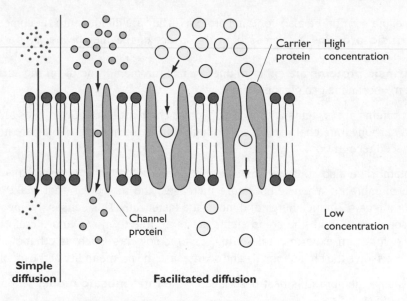

Figure 24 Diffusion and facilitated diffusion

The graphs in Figure 25 show the effect of difference in concentration across a membrane on the rate of diffusion and facilitated diffusion. As can be seen from the graph showing rate of diffusion, as the concentration difference between the inside and the outside of the cell increases the rate of diffusion increases. Increasing temperature, decreasing size and increasing solubility in lipid would also increase the rate of diffusion.

In facilitated diffusion, as in diffusion, as the concentration difference between the inside and the outside of the cell increases so does the rate of facilitated diffusion. However, at very high concentration differences the rate of facilitated diffusion reaches a maximum and levels off. This is due to the carrier proteins or channel proteins in the plasma membrane that are being used for facilitated diffusion becoming saturated, i.e. full all the time. Therefore further increases in the concentration difference

Knowledge check 24

How does diffusion differ from facilitated diffusion?

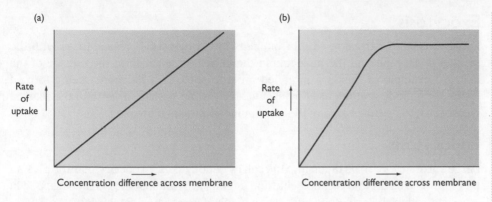

Figure 25 The effect of difference in concentration across a membrane on the rate of (a) diffusion and (b) facilitated diffusion

between the inside and the outside of the cell do not increase the rate of facilitated diffusion further.

Other factors that influence facilitated diffusion include temperature and number of carriers or channels in the membrane.

Active transport

Active transport is the movement of molecules or ions from a low concentration to a high concentration (Figure 26). As this movement is against the concentration gradient chemical energy in the form of ATP is required. A carrier protein in the plasma membrane is used as a pump. An example of active transport is the transport of mineral ions such as nitrates into the root hair cells of plants.

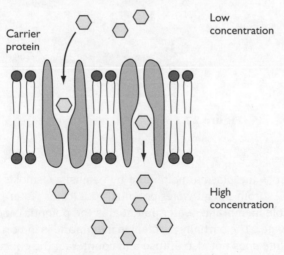

Figure 26 Active transport

In experimental conditions active transport can be stopped by the addition of a metabolic poison such as cyanide. Cyanide stops ATP being produced. As active transport requires ATP no active transport will occur.

Exocytosis

Large molecules can be released from cells by exocytosis. A vesicle fuses with the plasma membrane and the molecule in the vesicle is released to the outside of the cell. An example of this is a modified protein (such as a hormone) being formed in the Golgi body and released by exocytosis. As the vesicle fuses with the plasma membrane, the surface area of the plasma membrane is increased.

Endocytosis

Large substances can be taken into the cell by endocytosis. In endocytosis the plasma membrane folds around the molecule and engulfs it. The substance is then trapped in a vesicle within the cell. Endocytosis can be divided into two categories:

- **phagocytosis** — endocytosis of large solid substances, e.g. a white blood cell ingesting bacteria
- **pinocytosis** — endocytosis of fluids

As a vesicle is formed from the plasma membrane in endocytosis it decreases the surface area of the plasma membrane. Figure 27 illustrates the processes of endocytosis and exocytosis.

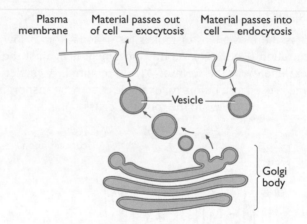

Figure 27 Endocytosis and exocytosis

Osmosis

Water moves across the plasma membrane by osmosis. Osmosis is the movement of water molecules from a high water potential to a lower water potential across a partially permeable membrane. Water potential is the potential energy of a solution relative to pure water. The partially permeable membrane is important in osmosis as it ensures the solute does not also diffuse and counteract the effect of osmosis.

water potential of a cell (Ψ_{cell}) = solute potential (Ψ_s) + pressure potential (Ψ_p)

The solute potential (Ψ_s) is generated by the solutes dissolved in the water.

The pressure potential (Ψ_p) is the pressure generated by the cytoplasm pushing on the cell wall of a plant cell. As the cell wall is rigid and inelastic it resists this pressure.

Pure water has a water potential of 0 kPa. This is the highest possible water potential so all solutions have a negative water potential. Osmosis can sometimes be described as movement from a region of more negative to one of less negative water potential. During osmosis the solution with the highest water potential is known as the **hypotonic** solution, while the solution with the lower potential is **hypertonic**. Osmosis will continue until both solutions have the same water potential. The two solutions will then be **isotonic**. At this point water will still move but there will be no net movement of water (so overall, the water potentials will remain the same).

Plant cells placed in a hypertonic solution will lose water and become flaccid and plasmolysed. Plasmolysis is where the cytoplasm shrinks and comes away from the cell wall. This will cause a plant to wilt. If plants cells are placed in a hypotonic solution, they will gain water, swell and become turgid. This is important as it ensures plants remain upright. Animal cells placed in a hypertonic solution will lose water and become shrivelled. If animal cells are placed in a hypotonic solution, they will gain water, swell and then burst (lyse).

Figure 28 summarises the behaviour of plant and animal cells in hypotonic, hypertonic and isotonic solutions.

Examiner tip

It is easy to get confused when working out directions of osmosis. Remember water always travels from a higher water potential (closer to zero) to a lower water potential (further from zero).

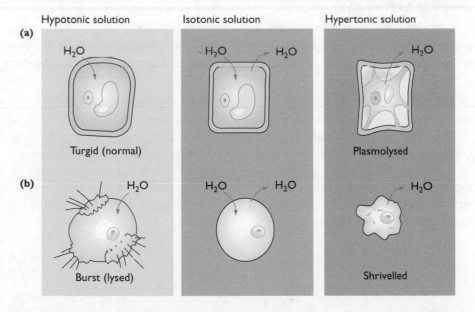

Figure 28 The behaviour of (a) plant cells and (b) animal cells in hypotonic, hypertonic and isotonic solutions

Knowledge check 25

What happens to a plant cell when it is placed in a hypertonic solution?

In plants incipient plasmolysis is the point at which the cytoplasm begins to come away from the cell wall. It can be found experimentally by observing the solute potential at which half of the plant cells in a sample are plasmolysed. At this point the solute potential is equal to the water potential. This is due to the pressure potential being zero as the cytoplasm is no longer pushing against the cell wall.

Summary

After studying this topic you should be able to:

- Explain the general functions of a cell's plasma membrane.
- Describe the structure of the plasma membrane in terms of the fluid mosaic model to include the phospholipid bilayer, intrinsic and extrinsic proteins, the function of carbohydrate chains and cholesterol.
- Explain the effect of organic solvents and temperature on the permeability of the plasma membrane.
- Describe the processes of diffusion, facilitated diffusion and active transport to include examples, direction of movement relative to concentration gradient, ATP requirement and route through the plasma membrane.
- Explain the processes of exocytosis and endocytosis.
- Describe osmosis as the movement of water only from a high water potential to a low water potential through a partially permeable membrane.
- Understand the water potential equation and be able to rearrange it to calculate the pressure and solute potentials.
- Use the terms hypotonic, hypertonic and isotonic to describe different solutions and be able to state the direction of water movement when given these solutions.
- Describe and explain the effect of placing animal cells and plant cells in hypotonic and hypertonic solutions.

Enzymes

Enzymes are biological catalysts. They speed up chemical reactions in living organisms and are unchanged at the end of the reactions. The fact they are unchanged is very important because it means they can be reused. Enzymes are specific — each one only catalyses one reaction.

In order for a reaction to occur a certain level of **activation energy** is required. Enzymes enable the reaction to occur with a lower activation energy. This means chemical reactions can occur at high rates even at relatively low temperatures, such as those found in the cells of living organisms. Activation energy is illustrated in Figure 29. Enzymes can either function **intracellularly** (within cells) or **extracellularly** (outside cells). Human digestive enzymes are an example of cells that work extracellularly.

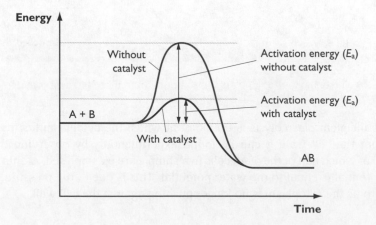

Knowledge check 26

How do enzymes speed up chemical reactions?

Figure 29 The effect of a catalyst on the activation energy of a reaction

Enzymes are **globular proteins**. They have a specific tertiary structure maintained by hydrogen, disulphide and ionic bonds. They have a region known as an **active site**. The substrate (the molecule or molecules that are being reacted) enters the active site and an **enzyme–substrate complex** forms. Products form and are then released from the active site. The enzyme is unchanged and is available to catalyse another reaction. This is shown in Figure 30.

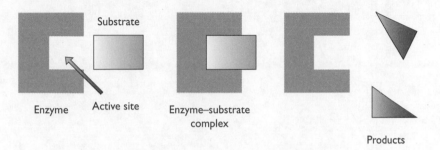

Figure 30 Enzyme action

In order for a product to be formed a successful collision (one with enough activation energy) has to occur between a substrate molecule and the active site of an enzyme.

There are two theories of how enzymes act:
- **The lock-and-key theory.** The enzyme's active site is a complementary shape to the substrate. The substrate enters the active site and an enzyme–substrate complex is formed. The products then form and leave the active site.
- **The induced-fit theory.** A more recent theory of enzyme action is induced fit. In this theory the active site slightly alters its shape to fit the substrate. Again the enzyme–substrate complex forms and products are released. The enzyme's active site then returns to its original shape. An example of an enzyme that is thought to work through the induced-fit theory is **lysozyme**. Lysozyme damages bacterial cell walls and can be found in tears and saliva.

Enzyme reaction graphs

Enzyme reactions are often studied by either measuring the concentration of the substrate over time or the concentration of product over time. This allows the rate of the reaction to be calculated. Exam questions on enzymes are often based around graphs that show the results of these investigations. On these graphs the rate of reaction is shown by the gradient of the line. A steeper gradient indicates a faster rate of reaction. Any factor that increases the number of successful collisions between active sites and substrates per unit time will increase the rate of reaction.

Knowledge check 27

What type of chemical are enzymes?

Examiner tip

The key aspect of an enzyme's action is its active site. Examiners will expect you to explain its importance and relate it to your knowledge of protein structure.

The graph in Figure 31 shows the concentration of products in a reaction over time. The reaction is occurring at a constant temperature and pH.

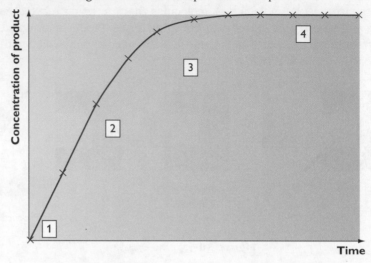

Figure 31 Product concentration over time

1 At zero time the concentration of products is zero as the reaction has not begun.
2 As time passes the concentration of products increases. At this stage the enzyme concentration is the limiting factor. If more enzymes were added the reaction would have a faster rate.
3 The reaction begins to slow down. This is due to most of the substrate being converted into products. The substrate is running out and there is therefore less chance of a successful collision between a substrate molecule and an active site. At this point the substrate concentration is the limiting factor in the reaction.
4 The concentration of products levels out and remains constant. All the substrate has been converted into product and the reaction has stopped.

This same reaction can be illustrated by a substrate concentration over time graph as shown in Figure 32.

Examiner tip
Students often score poorly on exam questions based on the graphs shown here. Do not just learn the general shapes. Ensure you fully understand what is happening at each stage of the graphs.

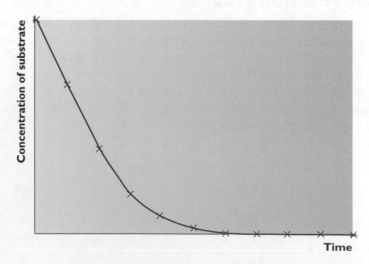

Figure 32 Substrate concentration over time

Again the rate of reaction is initially high (graph has a steep gradient) and is limited by the enzyme concentration. The rate of reaction then slows as most of the substrate is converted into product. At the end of the reaction the substrate concentration is zero as it has all been converted to product.

The use of a control is important in enzyme investigations. A control shows that it is the enzymes that are causing the effect that you are measuring (the dependent variable). The usual control in an enzyme reaction is to boil and cool the enzyme solution. This denatures the enzymes and ensures they will not catalyse any reactions. The experiment should then be repeated using the boiled and cooled enzyme solution. If the dependent variable still changes in the same way, this indicates that some other factor is influencing the results, not just the enzyme action.

As well as understanding the general effects of an enzyme you also need to learn how the following factors influence the rate of an enzyme-controlled reaction:
* temperature
* pH
* substrate concentration
* enzyme concentration
* competitive and non-competitive inhibitors

Temperature

Enzyme action is affected by temperature. Figure 33 shows the effect of increasing temperature on the rate of an enzyme-catalysed reaction.

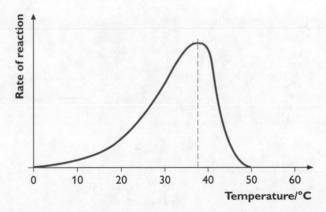

Figure 33 The effect of temperature on rate of an enzyme-catalysed reaction

At low temperatures the enzymes and the substrates have a low kinetic energy. This reduces the rate at which the molecules move within a solution. This means that there are fewer successful collisions between active sites and substrates. Fewer enzyme–substrate complexes are formed and therefore fewer products are formed per unit time. This means the rate of reaction is low at lower temperatures.

As the temperature increases the enzymes and the substrates gain more kinetic energy. This increases the rate of movement of molecules within the solution, increasing the chance of successful collisions between active sites and substrates. This leads to the formation of more products per unit time. Therefore the rate of reaction increases.

The rate of reaction continues to increase as the temperature increases until the enzyme's optimum temperature is reached. At the optimum temperature the rate of reaction is at its highest. At higher than the optimum temperature the rate of reaction decreases rapidly. This is because the enzyme's active sites become denatured. As the molecules gain kinetic energy they vibrate. Above the optimum temperature, these vibrations are so great that they cause the hydrogen bonds holding the enzyme's tertiary structure in place to break. This causes the active site to change shape and the enzyme is denatured. If the active site changes shape, the substrate can no longer bind with the enzyme so no enzyme–substrate complexes are formed and no products are formed.

As the temperature continues to increase more active sites denature. The rate of reaction continues to fall until all the active sites have denatured and the reaction stops.

Different enzymes have different optimum temperatures. Most human enzymes have optimum temperatures just below body temperature. This ensures that a small increase in the body's temperature does not denature the enzymes.

pH

As well as having an optimum temperature enzymes also have an optimum pH. Figure 34 shows the effect of pH on the rate of reaction of two enzymes: pepsin and amylase.

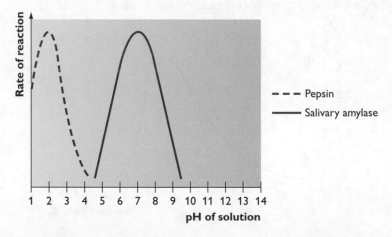

Figure 34 Rate of pepsin and salivary amylase-catalysed reactions at different pH values

The rate of the enzyme-catalysed reaction is highest at the optimum pH. If the pH of the solution either increases or decreases from the optimum, the rate of reaction falls:

- A small change from the optimum can cause the enzyme to become temporarily inactivated. This is due to changes in the charge of the active site causing the substrate to be repelled from the active site, preventing the formation of an enzyme–substrate complex. This is only a temporary change so if the enzyme is returned to its optimum pH the enzyme will again work as normal.
- A large change from the optimum will cause the enzyme's tertiary structure to change, altering the shape of the active site and resulting in the enzyme becoming permanently denatured.

Examiner tip

Not all of the enzymes in a reaction denature at just above the optimum temperature. As the temperature increases more of them will denature, hence the fall in the rate of reaction.

Examiner tip

Notice that the shapes for the temperature curve and the pH curve are different. The gradient of the pH curve is the same shape either side of the optimum as decreasing and increasing the pH has the same effect. The effect of high temperature is much more pronounced as the enzymes begin to denature above the optimum temperature.

As can be seen from Figure 34 different enzymes have a different optimum pH. Salivary amylase, a digestive enzyme found in saliva, has a slightly alkaline optimum pH while pepsin, a digestive enzyme found in the stomach, has an optimum pH of 2. This allows it to maintain its maximum rate of reaction in the acidic conditions found in the stomach.

As enzymes are sensitive to pH when carrying out an enzyme investigation it is important to ensure the pH remains constant. A pH buffer can be used to do this.

Substrate concentration

The concentration of the substrate also influences the rate of an enzyme-controlled reaction. The effect of increasing substrate concentration on the rate of reaction of an enzyme is shown in Figure 35.

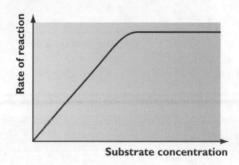

Figure 35 The effect of substrate concentration on the rate of an enzyme-catalysed reaction

As substrate concentration increases the rate of reaction also increases. This is because the presence of more substrate molecules increases the chance of successful collisions between active site and substrate. The rate of reaction continues to increase with increasing substrate concentration until a maximum rate of reaction is reached, after which further increases in substrate concentration no longer increase the rate of reaction.

A maximum rate of reaction is reached due to there being a fixed concentration of enzymes. At this point all the active sites of the enzymes are full all the time so the maximum number of successful collisions is occurring. This means there can be no increase in product formed per unit time. As it is the enzymes that are preventing the rate of reaction from increasing further we can say that the enzyme concentration is the limiting factor. This maximum rate of reaction can be increased if the concentration of enzymes is increased. This addition of enzymes will increase the number of available active sites. The rate of reaction would again continue to increase as substrate concentration increases. However, eventually the concentration of enzymes would again become the limiting factor.

Knowledge check 28

Why does pepsin have a much lower optimum pH than amylase?

Examiner tip

The graph showing the effect of increasing substrate concentration (Figure 35) is the same shape as the graph showing the change in concentration of product over time (Figure 31). When they come up in exam questions students often confuse these two graphs. Make sure you realise they are showing two separate things.

Examiner tip

When describing the point at which all the active sites are full all the time we can say that the graph has levelled off or plateaued.

Knowledge check 29

Why does the graph of rate of reaction over substrate concentration level off?

Enzyme concentration

Enzyme concentration also affects the rate of an enzyme reaction. Figure 36 shows the effect of enzyme concentration on the rate of an enzyme-controlled reaction.

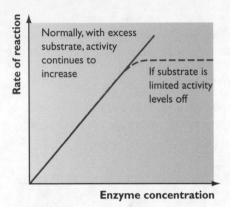

Figure 36 The effect of enzyme concentration on rate of reaction

As the enzyme concentration increases the rate of reaction also increases. This relationship only occurs if the substrate is in excess. This means that there is always enough substrate available to fill all the available active sites. This ensures the substrate concentration is not a limiting factor for the rate of reaction. If the substrate is not in excess then a maximum rate of reaction will be reached and the graph will level off.

Enzyme inhibitors

Inhibitors are molecules that can reduce the ability of an enzyme to speed up a reaction. There are two types: **competitive** and **non-competitive**.

Competitive inhibitors

Competitive inhibitors have a similar shape to the substrate molecule for an enzyme. This allows them to enter the active site of the enzyme and prevent the actual substrate binding. As the substrate cannot bind enzyme–substrate complexes cannot form, products are not formed and therefore the rate of reaction is not as high as it would be if the inhibitor were not present. An example of a competitive inhibitor is malonic acid. Figure 37 illustrates the action of a competitive inhibitor.

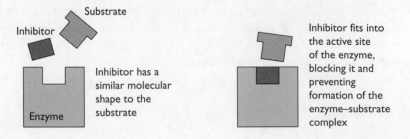

Figure 37 A competitive inhibitor competes with the substrate for the active site

Non-competitive inhibitors

Non-competitive inhibitors bind to an area of an enzyme other than the active site (e.g. an allosteric site). The inhibitor binding causes the active site of the enzyme to change shape. This means that the original substrate molecule can no longer bind to the active site. Examples of non-competitive inhibitors include cyanide and mercury. Figure 38 illustrates the action of a non-competitive inhibitor.

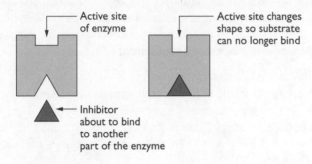

Figure 38 One type of non-competitive inhibitor

Knowledge check 30

How does the action of a competitive inhibitor differ from the action of a non-competitive inhibitor?

Figure 39 shows the effect of increasing substrate concentration on an enzyme-controlled reaction in the presence of no inhibitor, a competitive inhibitor and a non-competitive inhibitor.

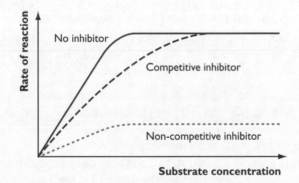

Figure 39 The effect of substrate concentration on rate of reaction in the presence of a competitive inhibitor and a non-competitive inhibitor

As can be seen from Figure 39, the effect of a competitive inhibitor can be counteracted by increasing the substrate concentration. As the substrate concentration increases the chance of a successful substrate and active site collision becomes much greater than the chance of an inhibitor and active site collision. Therefore, as the substrate concentration increases the relative effect of the inhibitor will be reduced. Eventually, if the substrate concentration is high enough the competitive inhibitor will no longer be able to affect the rate of reaction and the original theoretical maximum rate of reaction will be reached.

As in the presence of a competitive inhibitor, increasing the substrate concentration will counteract the effect of a non-competitive inhibitor and lead to the rate of reaction rising (as shown in Figure 39). However, the rate of reaction will never reach

Examiner tip

The concepts illustrated in Figure 39 are difficult to understand. The key point to realise is that the non-competitive inhibitor is reducing the number of active sites. As the number of active sites is the rate-determining factor at high substrate concentrations a reduction in active sites reduces the maximum rate of reaction.

Examiner tip

When answering questions on immobilised enzymes it is important not just to say 'can be reused' as an advantage. Enzymes are biological catalysts so they can all be reused. They key advantage to immobilised enzymes is they can be easily removed from the reaction solution in order to be reused.

the maximum, no-inhibitor-present level. This is because the non-competitive inhibitor reduces the number of available active sites in the reaction. Therefore no matter how much the substrate concentration is increased the rate of reaction will never reach its maximum.

Immobilised enzymes

Enzymes can be very useful in industrial processes, but some of their properties can make them inefficient to use. To overcome this problem the enzymes can be immobilised — they are fixed to an inert support or trapped in a matrix. Examples of supports include alginate beads and gel membranes.

Immobilising enzymes has several advantages:
- The enzymes are more stable at higher temperatures and a wider range of pH.
- The enzymes can be easily added or removed. This can be used to control the reaction and ensure a pure product is formed. It also means the enzymes can be easily recovered from the product. This allows the enzymes to be easily reused in the industrial process.
- A mixture of enzymes with different optimum pH and temperatures can be successfully used together.

Biosensors

Immobilised enzymes can be used in biosensors. Biosensors use a biological molecule to detect a particular substance. Immobilised enzymes are used in biosensors because they are specific to one substrate so they can accurately detect this substrate in a mixture. Immobilised enzymes are also able to detect low concentrations of a particular molecule in a mixture and they produce fast results.

Biosensors can be used to measure the concentration of glucose in the blood of a diabetic person, for example. The enzyme glucose oxidase is used and the process is outlined below:
- The blood sample is introduced into the biosensor.
- The glucose in the blood diffuses through the partially permeable membrane and enters the active site of the immobilised glucose oxidase.
- The reaction catalysed by the glucose oxidase causes oxygen to be taken up.
- The higher the concentration of glucose present, the more oxygen is taken up.
- The fall in oxygen concentration is detected by an oxygen electrode.
- The oxygen electrode acts as a transducer; it converts a chemical signal into an electrical signal.
- This electrical signal is processed and a readout is given, on a meter, of the glucose concentration in the blood.

After studying this topic you should be able to:
- Describe enzymes as biological catalysts that lower the activation energy of a reaction and are unchanged at the end of it.
- Explain the action of an enzyme and relate its specificity to the tertiary structure of its active site.
- Explain the differences between the lock-and-key and induced-fit theories of enzyme action.
- Interpret graphs of enzyme-catalysed reactions showing the change of product or substrate over time.
- Interpret graphs showing the effects of temperature, pH, substrate concentration and enzyme concentration on rate of reaction.
- Explain the effect of each of these factors using collision theory.
- Explain the action of competitive and non-competitive inhibitors and interpret graphs showing the effect of increasing substrate concentration on rate of reaction when inhibitors are present.
- Describe the importance of a control experiment and the use of pH buffers in enzyme investigations.
- Explain the advantages of immobilising enzymes.
- Describe the use of biosensors and explain the advantages of using immobilised enzymes in them.

Nucleic acids

DNA and RNA are both nucleic acids. **Deoxyribonucleic acid (DNA)** is the molecule that carries an organism's genetic code. The main function of DNA is replication and providing the code for protein synthesis. DNA replication ensures the DNA is copied accurately for cell division. When a cell divides a copy of the DNA is passed to both daughter cells. **Ribonucleic acid (RNA)** is another nucleic acid that is involved in the synthesis of proteins.

Examiner tip
It is okay to use DNA and RNA as abbreviations in exam answers.

Structure of nucleotides

Nucleic acids like DNA and RNA are polymers of nucleotides (Figure 40). Nucleotides are made up of the following three components:
- **pentose (5 carbon) sugar** — in DNA the pentose sugar is **deoxyribose** and in RNA it is **ribose**
- **phosphate**
- **organic nitrogenous base**

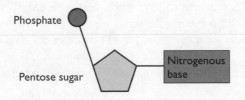

Figure 40 A generalised nucleotide

Structure of nucleic acids

Pyrimidine and purine bases

There are five different nitrogenous bases. They can be divided into two groups: pyrimidine bases and purine bases. Pyrimidine bases have a single-ring structure while purine bases have a double-ring structure. Figure 41 illustrates the different single- and double-ring structures of pyrimidine and purine bases.

- **Pyrimidine bases — single-ring structure:**
  - cytosine
  - thymine
  - uracil
- **Purine bases — double-ring structure:**
  - adenine
  - guanine

Examiner tip

Look out for the single- or double-ring structures of bases in DNA diagrams. It may help you to identify the base.

Knowledge check 31

What are the purine bases?

Knowledge check 32

What forms the backbone of a DNA strand?

Examiner tip

When answering questions about the structure of DNA it is a good idea to try to use the key term **complementary base pairing**. Also in an exam answer do not abbreviate the bases to their first letters, use their full names.

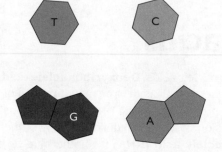

Figure 41 Comparison of the different structures of pyrimidine and purine bases

Adenine, thymine, cytosine and guanine are found in DNA.

Adenine, uracil, cytosine and guanine are found in RNA.

DNA

DNA is made up of two strands of nucleotides wound into a double helix (Figure 42). The deoxyribose of one nucleotide forms a bond with the phosphate of another nucleotide to form the **sugar–phosphate backbone**. The two strands of the double helix are held together by hydrogen bonds between pairs of nitrogenous bases. A purine base pairs with a pyrimidine base.

- Adenine always pairs with thymine. Three hydrogen bonds form between adenine and thymine.
- Cytosine always pairs with guanine. Two hydrogen bonds form between cytosine and guanine.

This is known as **complementary base pairing** and is important in ensuring DNA replicates correctly. It also means that there is the same proportion of adenine and thymine in a DNA molecule and also the same proportion of guanine and cytosine.

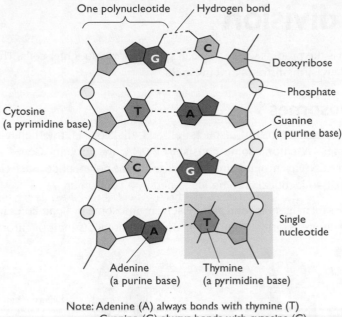

Note: Adenine (A) always bonds with thymine (T)
Guanine (G) always bonds with cytosine (C)

Figure 42 The molecular structure of DNA, showing the two antiparallel polynucleotides

RNA

There are three different types of RNA:

- **messenger RNA** (mRNA) is formed in the nucleus by transcription of a strand of DNA. mRNA carries the code for the polypeptide chain that will be formed in the translation stage of protein synthesis. After it has formed the mRNA leaves the nucleus through the nuclear pore.
- **ribosomal RNA** (rRNA) combines with protein to form ribosomes.
- **transfer RNA** (tRNA) transports amino acids to the ribosome for translation. tRNA has an anticodon and an amino attachment site.

Knowledge check 33

Where is mRNA formed?

After studying this topic, you should be able to:

- Describe the functions of DNA as replication in dividing cells and carrying the code for protein synthesis.
- Describe the structure of a nucleotide and the components that make it up.

- Explain that nucleotides form the polymers DNA and RNA.
- Understand the concept of complementary base pairing and use it to describe the structure of DNA.
- Describe the structural differences between DNA and RNA.

Summary

Examiner tip

Many similar sounding words are used in cell division (e.g. chromosome, chromatid, centromere, centriole). Make sure you are clear about what each one means.

Cell division

Cell division is the process whereby a cell splits to form daughter cells. There are two types of cell division: **mitosis** and **meiosis**.

Chromosomes

One of the most important aspects of cell division is the behaviour of the **chromosomes**. A chromosome consists of DNA and protein. Genes are specific sections of the DNA in chromosomes that code for one polypeptide. Each chromosome is made up of two identical chromatids joined by a centromere.

Chromosomes form homologous pairs. Each chromosome in a pair contains the same genes but in different forms. Figure 43 shows a homologous pair of chromosomes.

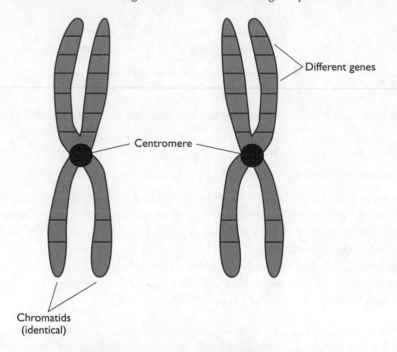

Figure 43 A homologous pair of chromosomes

A cell with the full number of chromosomes is known as **diploid**. In humans this means a cell contains 46 chromosomes forming 23 homologous pairs.

A cell with only one chromosome from each homologous pair (so half the diploid number) is known as **haploid**. Human haploid cells contain 23 chromosomes. Haploid cells are used as **gametes** by sexually reproducing organisms. When two haploid gametes fuse during fertilisation they form a cell which has the full diploid number of chromosomes.

Different organisms have different numbers of chromosomes, for example cows have 60 chromosomes while dogs have 78.

Knowledge check 34

What joins the two chromatids in a chromosome together?

Examiner tip

You do not need to learn the number of chromosomes of any particular animal other than humans, but when given examples you should be able to apply the concepts of haploid and diploid numbers of chromosomes to them.

Knowledge check 35

An organism has a diploid number of 54. How many chromosomes would there be in a haploid cell from the same organism?

Mitosis and the cell cycle

Although mitosis and meiosis both produce new daughter cells, the characteristics of these daughter cells are quite different:

- Mitosis produces two diploid daughter cells that are genetically identical to each other and to the parent cell. This provides genetic stability.
- Meiosis produces four haploid daughter cells that are genetically different to each other and to the original parent cell. The daughter cells formed in meiosis are used to produce gametes for sexual reproduction.

The cell cycle is shown in Figure 44.

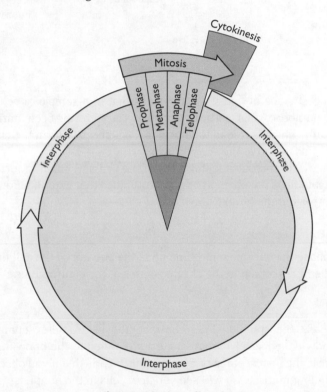

Figure 44 The cell cycle

When a cell is not undergoing cell division it is in **interphase**. During interphase the following processes occur:

- DNA replicates
- protein synthesis occurs
- ATP is synthesised
- organelles are produced

At the end of interphase mitosis occurs. At the end of mitosis the cell divides by **cytokinesis** to form two genetically identical daughter cells.

Mitosis can be divided into four stages. **Prophase, metaphase, anaphase** and **telophase** (Figure 45). It is important to realise that mitosis is a continuous process and we divide it up into separate stages to aid our understanding of it.

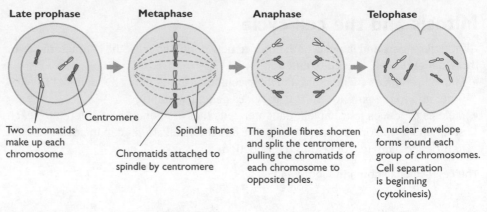

Figure 45 The stages of mitosis

Prophase

This is the longest stage of mitosis. This means that in a sample of cells undergoing mitosis a large number of cells will be in prophase. The following occur during this stage:
- The chromatin, which is free in the nucleus, condenses to form the chromosomes, (two chromatids joined at the centromere).
- Centrioles move to the poles (either end) of the cell and begin to form the spindle fibres. The spindle is a web of protein microtubules that extend across the cell.
- The nuclear membrane breaks down.

Metaphase

During metaphase the chromosomes line up at the equator of the cell (in the middle) and the spindle fibres attach to the chromosomes at the centromeres.

Anaphase

During anaphase the spindle fibres contract, the centromere separates and the chromatids (now called sister chromosomes) are pulled to the opposite poles of the cell. Anaphase is the fastest stage of mitosis. This means that if you observe a sample of cells undergoing mitosis you will see very few cells in the anaphase stage.

Telophase

During telophase the chromosomes unwind back to chromatin. The nuclear envelope reforms and the spindle breaks down.

Cytokinesis

At the end of telophase the cell divides by **cytokinesis** to form two daughter cells. In animal cells this is done by a 'pinching in' of the plasma membrane. In plant cells a cell plate forms between the dividing cells. This then forms the new cell wall of the two cells.

Functions of mitosis

Mitosis is important for the growth of organisms, the repair of damaged tissues and the replacement of dead cells, for example skin cells. Some organisms also use mitosis for asexual reproduction. Examples of organisms that use mitosis to reproduce asexually include bulb-producing plants, such as tulips, and fungi, such as yeast. Uncontrolled mitosis leads to the formation of groups of cells called **tumours**. Benign tumours are very slow growing and are mostly harmless. Malignant tumours are cancerous and can spread into other tissues and organs.

Meiosis

In sexually reproducing organisms meiosis is used to form gametes. As mentioned previously, gametes are haploid cells which contain half the diploid number of chromosomes.

Meiosis has two stages **meiosis I** and **meiosis II**. You do not need to know the full detail of meiosis but you do need to know how it differs from mitosis. The key difference is that meiosis produces four haploid daughter cells that are genetically different to each other and to the original parent cell. This genetic variation in gametes leads to the genetic variation seen in the offspring of sexually reproducing organisms.

Meiosis I

Meiosis I is divided into four stages: prophase I, metaphase I, anaphase I and telophase I.

As in prophase of mitosis, during prophase I the nuclear membrane breaks down, the centrioles move to the opposite poles of the cell and form the spindle. The chromatin also condenses to form the chromosomes. In meiosis I the homologous chromosomes pair up to form bivalents (this does not occur in mitosis).

Chromosomes in a bivalent are able to swap genes in a process called **crossing over**. This crossing over occurs at points where the homologous chromosomes touch — these points are called **chiasmata**. Crossing over is one of the sources of variation in meiosis. Figure 46 illustrates the process of crossing over.

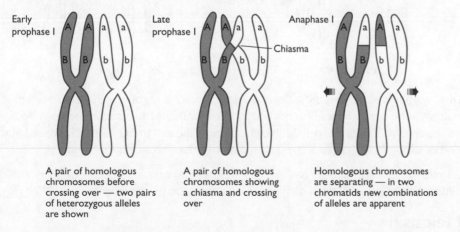

A pair of homologous chromosomes before crossing over — two pairs of heterozygous alleles are shown

A pair of homologous chromosomes showing a chiasma and crossing over

Homologous chromosomes are separating — in two chromatids new combinations of alleles are apparent

Figure 46 Crossing over and genetic variation

During metaphase I the bivalents line up at the equator of the cell and spindle fibres join to the centromere of one chromosome in each bivalent. The bivalents arrange themselves randomly at the equator. This random distribution of chromosomes leads to **independent assortment**. This means that the resulting daughter cells will receive a mix of chromosomes. This is another source of genetic variation in meiosis. Independent assortment is illustrated in Figure 47.

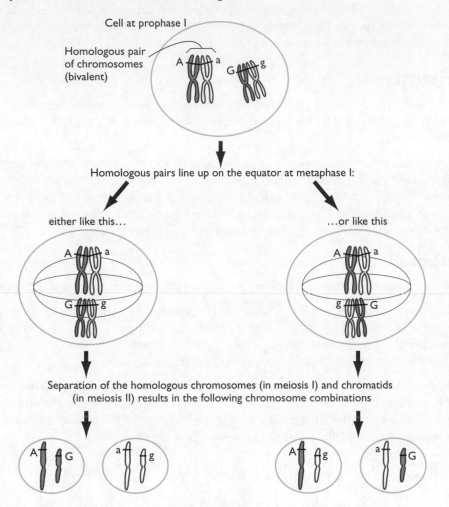

Figure 47 Independent assortment and genetic variation

During anaphase I the spindle fibres contract and one chromosome from each pair is pulled to the opposite pole of the cell. Unlike in mitosis the centromere does not split so the chromosomes are still double structures. The cell then enters telophase I at the end of which cytokinesis occurs producing two daughter cells.

As each daughter cell only has one of each pair of homologous chromosomes these cells are now haploid. Both daughter cells now undergo meiosis II.

Meiosis II

Meiosis II is similar to mitosis. Meiosis II ends with four daughter cells produced, each with half the original number of chromosomes (haploid) and with each daughter cell being genetically different to each other and to the parent cell.

When these haploid gametes fuse during fertilisation the chromosomes of both parents are mixed. This leads to further genetic variation in the offspring.

Sources of variation

In summary, there are three main sources of variation in meiosis:

- crossing over — during prophase I homologous chromosomes in a bivalent swap genes at points called chiasmata
- independent assortment — during metaphase I homologous chromosomes in bivalents arrange themselves randomly at the equator
- mixing of parental chromosomes at fertilisation

After studying this topic, you should be able to:
- Understand the relationship between chromatin, chromatids and homologous chromosomes.
- Describe the processes that occur in interphase.
- Describe the events that occur in each of the stages of mitosis.
- Explain the differences between mitosis and meiosis.
- Describe the sources of genetic variation in meiosis.

Summary

Questions & Answers

This section contains questions on each of the topic areas in the specification. They are written in the same style as the questions in the BY1 exam so they will give you an idea of the sort of thing you will be asked to do in the exam. After each question there are answers by two different students followed by examiner's comments on what they have written. These are important because they give you an insight into the responses the examiners are looking for in the exam. They also highlight some of the common mistakes students make.

The AS Unit I paper

The BY1 examination lasts 1 hour 30 minutes and is worth 70 marks. The first 60 marks are for answering structured questions and the last 10 marks are for writing an essay. In the exam you will be given a choice of two essays questions, but you only need to answer one of them.

In addition to sample structured questions, this section of the guide also contains two example essays along with some advice on how best to approach essay questions in general.

Examiner's comments

Examiner comments on the questions are preceded by the icon ⓔ. They offer tips on what you need to do in order to gain full marks. All student responses are followed by examiner's comments, indicated by the icon ⓔ, which highlight where credit is due. In the weaker answers, they also point out areas for improvement, specific problems and common errors such as lack of clarity, irrelevance, misinterpretation of the question and mistaken meanings of terms.

Biochemistry

Question 1

(a) Identify the two monosaccharides shown below. (2 marks)

(b) Describe one way in which the structure of the monosaccharide found in DNA would differ from A and B. (1 mark)

(c) Describe the formation of maltose from two monosaccharides (4 marks)

(A)
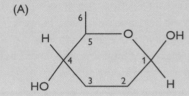

(B)
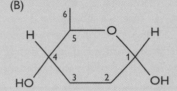

ⓔ This is a fairly straightforward biochemistry question, but it has a couple of areas that students could find challenging. It is important to study all biochemistry diagrams carefully as it is easy to get different molecules mixed up. Part (c) carries 4 marks so it is important to ensure you write a detailed answer that contains all the main points and key words required.

Student A

(a) A   Beta glucose
     B   Alpha glucose
(b) The monosaccharide found in DNA is deoxyribose while the monosaccharides shown in the diagrams are glucose.
(c) Two glucose molecules join together in a condensation reaction.

ⓔ **4/7 marks awarded** (a) Both the monosaccharides are identified correctly so Student A gains both the marks available. (b) Student A correctly identifies the monosaccharide in DNA as being deoxyribose, but this is not what the question was asking — the question was asking for the differences in structure between the monosaccharides shown in the diagram and the monosaccharide in DNA. The most obvious difference in structure is that the monosaccharides shown are isomers of glucose so are hexose sugars and contain six carbon atoms; deoxyribose is a pentose sugar so contains five carbon atoms — explaining this would have gained the mark. (c) Student A's answer to this question is correct and gains 2 marks: 1 mark for saying that maltose is formed from two glucose molecules and 1 mark for correctly identifying the reaction as a condensation reaction. However, a simple one-line answer like this will not score 4 marks. It is vital that you look carefully at the mark allocation of each question and ensure that in a question with a high mark allocation like this one you include enough detail to earn all the available marks.

Student B

(a) A Alpha glucose
   B Beta glucose

(b) The monosaccharide found in DNA is deoxyribose which is a pentose sugar so has five carbon atoms. The two molecules shown are hexose sugars so contain six carbon atoms.

(c) In order to form maltose two glucose molecules combine in a condensation reaction. The glycosidic bond in the molecule is formed between carbon one of one of the glucose molecules and carbon four of the other glucose molecule. A molecule of water is also formed.

🅔 **5/7 marks awarded** Overall, Student B answered this question well but did not score full marks owing to a silly mistake made in part (a). Most diagrams you will see of the two isomers of glucose will show alpha glucose first and beta glucose second. However, examiners often present things in non-standard ways to check that you have understood and thought about the topic. Make sure you study all diagrams carefully — if Student B had done this he or she would have realised that the arrangement of the hydrogen atom and OH group on carbon 1 in (A) shows that it is beta glucose. (b) This is an excellent answer that gains the mark. (c) This is a detailed answer that gets all 4 marks. Longer answer questions such as this one will often have more acceptable answers than there are marks available, for example for (c) there were five possible points. In this case the student got all five.

Question 2

(a) Give one structural similarity and one structural difference between amylose and amylopectin. (2 marks)

(b) What is the monomer of the polymer cellulose? (1 mark)

(c) Give a function of cellulose. (1 mark)

(d) What property of cellulose fibres makes them so suitable for this function? (3 marks)

🅔 It is important to take notice of the keywords in questions. Part (a) requires you to talk about structural differences. As it is asking for a similarity and difference you need to compare amylose and amylopectin, not just describe the characteristics of amylose or amylopectin.

(a) Amylose and amylopectin make up starch; amylose has different bonds to amylopectin.
(b) Beta glucose.
(c) Makes up cell wall.
(d) Long chains of beta glucose monomers are linked by hydrogen bonds to form microfibrils and many microfibrils link together to form cellulose fibres. This makes cellulose fibres very strong.

🅔 **4/7 marks awarded** (a) This answer gains no marks — the first point is not a reference to structure and 'different bonds' is not detailed enough to earn a mark. Reference should have been made to the 1-4 glycosidic bonds in amylose and the 1-4 and 1-6 glycosidic bonds in amylopectin. A mark could also have been gained by stating that amylose is a helix while amylopectin has a branched structure. (b) This answer is correct and gains 1 mark. (c) This answer lacks detail. Cellulose forms the cell walls of plants; other organisms have cell walls formed from different substances, e.g. the cell wall of bacteria is formed from murein. (d) This is a good answer that scores all 3 marks.

(a) Both amylose and amylopectin are polymers of alpha glucose. A difference between amylose and amylopectin is that amylose has 1-4 glycosidic bonds while amylopectin has 1-4 and 1-6 glycosidic bonds.
(b) Glucose.
(c) Forms plant cell walls.
(d) There are a large number of hydrogen bonds between the OH groups of beta glucose monomers that make up the cellulose chains in cellulose fibres. This makes them very strong.

🅔 **5/7 marks awarded** (a) This is an excellent, detailed answer that gains 2 marks. (b) This answer lacks detail so gains no marks. The monomer of cellulose is beta glucose. (c) This is a correct answer and gains 1 mark. (d) This is not a very detailed answer and only just scores 2 marks. A full answer would include detail of how the beta glucose monomers in the cellulose chains are rotated 180 degrees from the previous molecule. You could also give an explanation of how the cellulose chains form microfibrils and many microfibrils form a cellulose fibre.

Question 3

(a) Arginine and glycine are two different amino acids. How do arginine and glycine differ from each other? (1 mark)

(b) Label the bonds shown at A, B, C in the diagram below. (3 marks)

(c) Is the protein below a fibrous or globular protein? Explain your answer. (3 marks)

(d) How would a quaternary protein differ from the one shown below? (1 mark)

(e) What biochemical test could be used to determine if a protein is present in a solution? (1 mark)

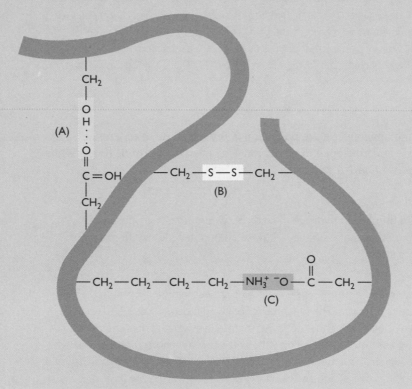

ⓔ Make sure you study any diagrams in the exam paper carefully. This question asks you not only to specifically identify features of the diagram, but also to look at it as a whole to identify the class of protein shown and compare it to a quaternary protein.

Student A

(a) Different structures.
(b) A hydrogen bonds, B disulphide bridges, C ionic bonds.
(c) The protein shown is a globular protein.
(d) A quaternary protein would be made up of two or more polypeptides. The protein shown only has one polypeptide chain.
(e) Add Benedict's reagent.

⊖ 5/9 marks awarded (a) This answer does not contain enough detail — to be awarded the mark the student needed to mention the different R groups of the two amino acids. (b) All three bonds are correct so this earns 3 marks. (c) Student A has correctly identified the protein in the diagram as a globular protein so gains 1 mark. However, the student has failed to answer the second part of this question and has not explained why the protein shown is a globular protein. It is really important when answering exam questions to take time to read each question fully and ensure your answer addresses all the points asked for. (d) This is a good answer and gains the mark. (e) The student has mixed up the biochemical test for reducing sugars and the test for protein. This is a common mistake (due to them both beginning with the letter 'B') and underlines the importance of learning the biochemical tests thoroughly.

Student B

(a) They would have different r groups.
(b) A hydrogen bonds, B ?, C ionic bonds.
(c) This is a globular protein, you can tell this as it has a tertiary structure. Fibrous proteins do not have a tertiary structure.
(d) A quaternary protein would be haemoglobin.
(e) The biuret test.

⊖ 7/9 marks awarded (a) This is the correct answer and gains the 1 mark available. (b) A and C are identified correctly, gaining 2 marks, but Student B has failed to recognise the disulphide bridge (can also be labelled as a covalent bond). It is important that you can recall facts like all the bonds in the tertiary structure of a protein as they are a source of easy marks. Student B should have been able to identify bond B from the S (chemical symbol for sulphur). (c) Student B has correctly identified the protein in the diagram as a globular protein and explained the structural features that distinguish it from a fibrous protein. (d) Student B has given an example of a quaternary protein without describing how it differs from the protein in the diagram so did not get the mark. (e) Student B has correctly identified the biochemical test for proteins so gains the mark.

Question 4

A student was given four solutions labelled A, B, C and D. She knew that the solutions contained:
- **Glucose**
- **Starch**
- **An enzyme that breaks down starch into glucose**
- **Sucrose**

The diagram below shows the test method she carried out:

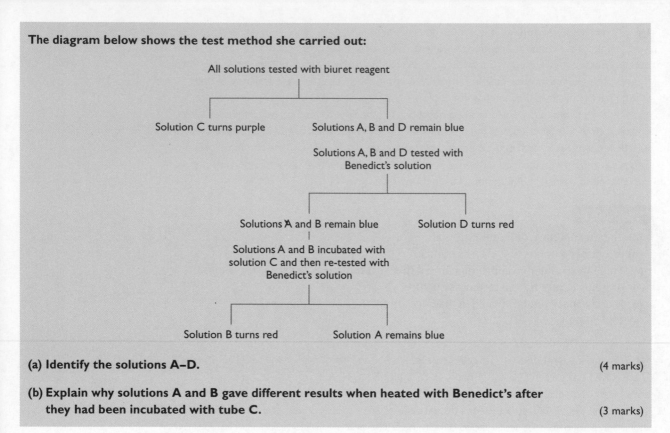

(a) Identify the solutions A–D. (4 marks)

(b) Explain why solutions A and B gave different results when heated with Benedict's after they had been incubated with tube C. (3 marks)

ⓔ This question is quite challenging. It is important to take your time and follow the experimental process through carefully. It may be helpful to write on the diagram when you think you have identified one of the solutions and then tick off that word in the stem of the question. Once you think you have worked out what each solution is go back through the question and ensure there is no evidence that contradicts your assumptions.

Student A

(a) A Glucose
B Starch
C Enzyme
D Sucrose
(b) The starch in solution B is broken down by the enzyme in solution C, the glucose produced then gives a positive result to being heated with Benedict's reagent.

ⓔ **3/7 marks awarded** (a) The student correctly identifies solutions B and C but has mixed up solutions A and D, so only scores 2 of the 4 marks available. (b) This answer is not explained well, but it is a good example of how you can still get some marks from a question even if you do not fully understand it. Student A has used his/her answer in part (a) and the information in the stem of the question to gain 1 mark out of the possible 3 available by stating the effect of the enzyme. If Student A had read back over the question and applied the logic of his/her answer in (b) to answering (a) he or she would have realised the mistake made (glucose would have given a positive result to the Benedict's test) and could have scored an extra 2 marks. This underlines the importance of always reading back over the answers you have written in an exam.

(a) A   Sucrose

B   Starch

C   Enzyme

D   Glucose

(b) The enzyme in tube C breaks down the starch into glucose, the glucose produced then gives a positive result to being heated with Benedict's reagent. As glucose is a reducing sugar it forms a brick red precipitate. Sucrose is not broken down by the enzyme and as it is a non-reducing sugar it does not form a brick red precipitate when heated with Benedict's reagent.

ℯ **7/7 marks awarded** (a) All the solutions are correctly identified, scoring all 4 marks. (b) This is an excellent answer that fully explains the reasoning behind the different results of solutions A and B. It scores the maximum 3 marks for: describing the effect of the enzyme; identifying the products of the enzyme reaction as being reducing sugars so producing a positive result when heated with Benedict's reagent; stating the fact that as sucrose is a non-reducing sugar it gives a negative result when heated with Benedict's reagent.

Cell structure

Question 5

The diagram below shows a prokaryotic cell.

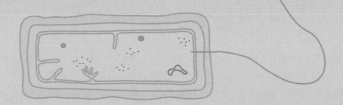

(a) Describe three features shown on the diagram that allow you to identify the above cell as a prokaryote. (3 marks)

(b) How would the cell wall in the above organism differ from that in a plant cell? (2 marks)

Ribosomes are found in both eukaryotes and prokaryotes.

(c) What is the function of ribosomes? (1 mark)

(d) Name one feature that is shown in the diagram above that would also be found in a virus. Name another feature that is found in viruses but is not found in prokaryotes or eukaryotes. (2 marks)

ⓔ This question is all about prokaryotes and viruses. Students often neglect this area of the cell structure/organisation topic and instead just focus on animal and plant cells. Make sure you have a good understanding of this area — focus especially on comparisons between eukaryotes and prokaryotes.

Student A

(a) Has a cell wall, DNA and flagella.
(b) The prokaryote cell wall is made of murein and a plant cell wall is made from cellulose.
(c) Protein synthesis.
(d) Viruses could have DNA which is shown above and a capsule, which eukaryotes and prokaryotes do not.

ⓔ **4/8 marks awarded** (a) This answer scores none of the 3 marks available. None of the features mentioned are specific to prokaryotes. Cell walls are also found in plant cells, DNA is found in all cells and flagella are found in a number of non-prokaryote cells such as sperm. (b) This is a good answer that gains both marks. (c) Protein synthesis is the correct function for ribosomes so gains the 1 mark available. (d) Student A has made a common mistake here and does not score 1 of the 2 marks available. Viruses have capsids while prokaryotes have capsules. DNA is found in both prokaryotes and some viruses though so this comparison does score 1 mark.

Student B

(a) The cell shown is a prokaryote cell as it has no membrane-bound organelles such as mitochondria, its DNA is free in the cytoplasm and it has a mesosome.
(b) The prokaryote cell wall is not made of cellulose.
(c) Protein synthesis.
(d) Viruses have a protein coat and eukaryotes and prokaryotes do not. Viruses could have DNA which is shown in the diagram above.

ⓔ **7/8 marks awarded** (a) This is a good, detailed answer that identifies three key prokaryote features shown in the diagram — it gains the full 3 marks. (b) This answer is correct, but as the question is worth 2 marks more detail is required. Always look carefully at the mark allocation for questions and ensure you make at least as many separate points as there are marks. The second mark would be awarded for giving murein as the chemical prokaryote cell walls are made from. (c) This is the correct answer and scores the 1 mark available. (d) This is a correct answer and scores both the available marks. Capsid instead of protein coat would also have been accepted.

Question 6

The photo below shows an electron micrograph of a eukaryotic cell.

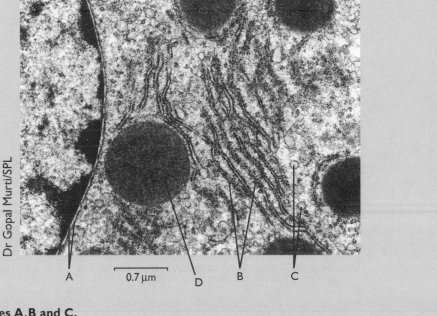

Dr Gopal Murti/SPL

A   0.7 µm   D   B   C

(a) Name the features A, B and C.   (3 marks)

(b) Structure A has pores along its length. Explain the function of these pores.   (1 mark)

(c) D is a mitochondrion. Give three ways in which the structure of this organelle is similar
to a chloroplast.   (3 marks)

ℯ Students often find questions with electron micrographs challenging. It is important to
remember that you will have learned about all the structures you will be asked to identify. If the
structure does not look immediately familiar try to think about the key features of organelles: what
general shapes are they, do they have internal structures etc.?

Student A

(a) A Nuclear membrane
    B Smooth endoplasmic reticulum
    C Vesicle
(b) Allows things to leave the nucleus.
(c) A chloroplast also has a double membrane, its own ribosomes and is used in
    respiration like a mitochondrion.

ⓔ **4/7 marks awarded** (a) Student A has correctly identified A and C, but B is the rough endoplasmic reticulum not the smooth endoplasmic reticulum. The distinguishing feature is the dots that can be seen on the endoplasmic reticulum in the diagram. These are ribosomes and are found on the rough endoplasmic reticulum. Student A therefore scores 2 of the 3 marks available. (b) This answer lacks sufficient detail to get the mark. The student should have stated what things leave the nucleus, for example mRNA. (c) This answer gains 2 marks for the first two points. However, the last point is not only wrong (chloroplasts carry out photosynthesis not respiration) but also the question asked for structural similarities. What the organelle does is a functional feature not a structural feature.

Student B

(a) A   Nuclear membrane
       B   RER
       C   Vesicle
(b) Allows mRNA to pass out of the nucleus into the cytoplasm.
(c) Like a mitochondria, a chloroplast has a double membrane, its own DNA and
        folded internal membranes that give a large surface area.

ⓔ **6/7 marks awarded** (a) Student B has correctly identified A and C but loses the mark for B for abbreviating rough endoplasmic reticulum to RER . With the exception of abbreviations of biochemicals such as DNA, RNA and ATP you should avoid abbreviations in exam answers. (b) This is an excellent answer that gains the mark. (c) This is a detailed, well-explained answer that gets all 3 marks.

Cell membranes and transport

Question 7

The diagram below shows a section of the plasma membrane.

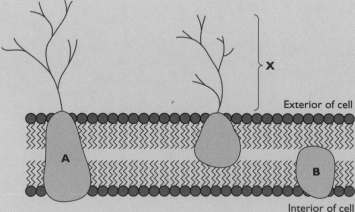

(a) Distinguish between:
   A and B
   X and B (2 marks)

(b) Give a possible function for X. (1 mark)

(c) The table below shows the rate of uptake of different substances into an animal cell.

Molecule	Rate of uptake into the cell/arbitrary units
Glucose	50
Na^+	10
Oxygen	1500
Carbon dioxide	900

How would the movement of glucose through the plasma membrane differ from the movement of oxygen through the plasma membrane? Explain this difference. (4 marks)

(d) Further investigation showed Na^+ ions were entering the cell against the concentration gradient. Explain how this is possible. (2 marks)

🄔 This question is all about comparisons. Think carefully about each aspect of the question and try to find the key differences. Ensure your answers contain both sides of the comparison.

Student A

(a) A is an intrinsic protein while B is an extrinsic protein. X is part of a glycoprotein and B is not.
(b) Cell–cell recognition.
(c) Glucose is a large, polar molecule so moves through the membrane by facilitated diffusion. This is through a carrier protein and requires energy in the form of ATP. Oxygen is a small non-polar molecule so can diffuse through the phospholipid bilayer.
(d) Nitrates are entering the cell by active transport.

🄔 **6/9marks awarded** (a) The first comparison is correct and scores 1 mark. The second comparison does not score the other available mark — X is a specific part of the glycoprotein (the carbohydrate chain) and Student A certainly will not score a mark for simply saying 'B is not' (b) This is the correct answer and scores the mark available; used in cell–cell signalling would also have been accepted. (c) Student A scores 3 of the 4 available marks for this question. He or she has correctly identified that glucose moves through the cell membrane by facilitated diffusion using a carrier protein, but has then stated that ATP is involved. This is incorrect and negates the mark that would have been awarded for explaining how glucose moves through the cell membrane. It is important point to remember that in the short-answer sections of the paper an incorrect statement among correct statements will cause you to lose a mark. (d) This answer is correct, but only scores 1 mark out of the 2 available as Student A has not explained how active transport can occur against the concentration gradient (by using ATP and a carrier protein as a pump).

Student B

(a) A is an intrinsic protein and B is an extrinsic protein. X is a carbohydrate and B is a protein.

(b) Cell–cell signalling.

(c) Glucose would move through the membrane by facilitated diffusion. Oxygen can diffuse through the phospholipid bilayer as it is a small lipid-soluble molecule.

(d) As nitrates are entering the cell against the concentration, active transport must be occurring. Active transport requires ATP and a carrier protein to act as a pump.

ⓔ **8/9 marks awarded** (a) This is a good comparison that scores both marks. (b) This is the correct answer and gets 1 mark. (c) Everything Student B has written for this answer is correct, but he or she has failed to explain why glucose has to move through the plasma membrane by facilitated diffusion (it is due to it being a large, polar molecule) so only scores 3 of the 4 marks available. (d) This is a well-explained answer that scores both available marks.

Question 8

The graph below shows the effect of lipid solubility on the rate of diffusion of molecules through a plasma membrane at two temperatures, 20°C and 50°C.

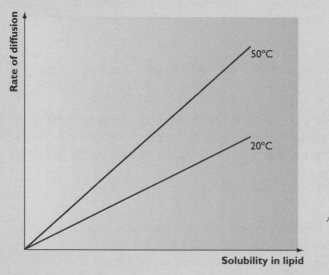

(a) Describe and explain the effect of lipid solubility of a molecule on the rate of diffusion through a cell membrane at 20°C. (2 marks)

(b) Describe and explain the difference between the results at 20°C and 50°C in terms of plasma membrane fluidity. (2 marks)

(c) The presence of which molecule in the plasma membrane reduces the fluidity of the plasma membrane? (1 mark)

e Scoring highly on this question depends entirely on fully understanding the graph. Take time to study it carefully and do not rush into answering the questions. Parts (a) and (b) both ask you to describe and explain. The fact that the question asks you to describe as well as explain means there will be marks available for just describing the trend of the graph. Even without a full understanding of the scientific theory it should be possible for all students to score some of the marks in this question. This illustrates the importance of not becoming disheartened with a question you are having difficulty understanding —there may still be some marks you can get. This question is a particularly good example of this as part (c) is just factual recall.

Student A

(a) When the solubility of the molecule in lipids increases it is able to move through the phospholipid bilayer of the plasma membrane much more easily.

(b) At 20°C the rate of diffusion is much less than at 50°C at all lipid solubilities. This is due to the diffusing molecules having a greater kinetic energy and therefore passing through the phospholipid bilayer much more easily.

(c) Cholesterol

e **2/5 marks awarded** (a) This answer scores 1 mark out of the 2 available. Student A has identified the link between lipid solubility and ability to pass through the phospholipids, but has failed to answer the describe part of the question. A simple statement saying that as the solubility in lipids increases the rate of diffusion through the plasma membrane increases would have scored the other mark. This is a classic example of a student answering the hard part of the question and assuming that in doing so he or she would also get the mark for the simpler describe section. It is vital that you clearly state the answers to the questions. In this case it is clear to the examiner that the student knows what the graph shows but unless the student states it the examiner cannot award the mark. (b) This statement is correct but scores neither of the 2 marks available. An increase in temperature would increase the kinetic energy of the diffusing molecules and so lead to a greater rate of diffusion, but the question specifically asks you to explain the difference in terms of the membrane's fluidity. It is a common mistake for students to answer the question they wished to be asked as opposed to the question that is asked. If the question specifically states that the answer should relate to the fluidity of the membrane then an answer that does not mention membrane fluidity will not get the marks. (c) This is the correct answer and gains 1 mark.

Student B

(a) As the solubility in lipid increases the rate of diffusion also increases. This is due to molecules which are lipid soluble being able to pass through the phospholipid bilayer of the plasma membrane more easily.

(b) At 50°C the components of the plasma membrane have more kinetic energy, therefore they move more causing the membrane to become more fluid. A more fluid membrane means that the molecules can pass through the phospholipid bilayer more easily, leading to the greater rate of diffusion shown at 50°C when compared to 20°C.

(c) Cholesterol

e **5/5 marks awarded** (a) This is an excellent answer which concisely describes and explains the trend of the graph and so scores both available marks. (b) This answer relates the difference between rates of diffusion at 20°C and 50°C to the increased fluidity of the membrane and explains this is due to an increase in kinetic energy of the components of the plasma membrane, scoring 2 marks. (c) This answer is correct and scores the 1 mark available.

Unit BY1: Basic Biochemistry and Organisation

Question 9

The diagram below shows the water potential of three plant cells

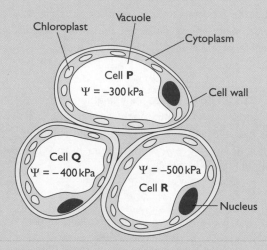

(a) Draw arrows to show the movement of water on the diagram above. (2 marks)

The photo below shows plant cells that have been put in two different concentrations of sucrose solutions, P and Q.

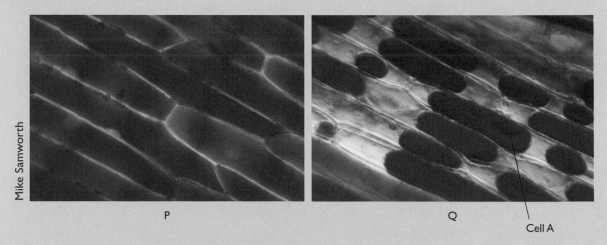

P   Q

Cell A

(b) Which of the solutions had the highest concentration of sucrose? Explain your answer. (2 marks)

(c) The solute potential of the cells in solution Q is −200 kPa. What is the water potential of cell A? Explain how you arrived at your answer. (4 marks)

ⓔ It is easy to make mistakes in water potential questions. Do not rush into an answer. Go through each part of the question methodically and always check back on the work you have done. Question (c) is particularly challenging. If you get to a question like this remember that the answer must be based on something in the specification, take a bit of time to think around the question and see if you can work out how it relates to what you know.

Student A

(a)

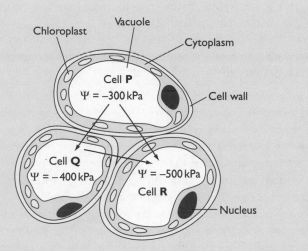

(b) Solution Q has the highest water potential of the two solutions. You can tell from the appearance of the cells.

(c) The water potential of the cell is 0 kPa as the cells are plasmolysed.

ⓔ **4/8 marks awarded** (a) The direction of movement of the water is correctly shown, going from a higher (less negative) water potential to a lower (more negative) water potential, and gets 2 marks. (b) Student A has correctly identified solution P as having the highest concentration of sucrose. However, he or she has failed to explain why so only scores 1 mark. (c) Student A clearly does not know how to answer this question. However, the student still scores 1 mark by using the keyword plasmolysed. This shows how it is possible to pick up marks even in complex questions by using appropriate keywords.

Student B

(a)

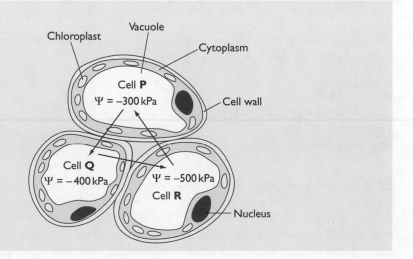

(b) Solution Q has the highest water potential of the two solutions as the cells in solution Q are plasmolysed, therefore water must have left the cell by osmosis, from a high to a low water potential.

(c) The water potential of the cell is –200 kPa. This is the same as the solute potential. As the cell is plasmolysed the pressure potential is zero. This is because the cytoplasm is not pushing against the cell wall.

℮ **7/8 marks awarded** (a) One of the arrows is pointing in the wrong direction. Student B therefore loses one of the available marks and only scores 1 mark. This incorrect answer to a simple question by an able student highlights the importance of checking answers involving water potential. The use of negative numbers can make it easy to make errors. (b) This is a detailed answer that scores both available marks. It would also be possible to score marks by explaining why solution Q does not have the highest concentration of sucrose — the water potential of this solution must be high as water moves into the cell so it is turgid. (c) Student B has correctly realised that as the cell is plasmolysed there is no pressure potential so therefore the solute potential is equal to the water potential. The student therefore scores all 4 of the available marks.

Enzymes

Question 10

(a) Why are enzymes known as 'biological catalysts'?   (2 marks)

(b) How is the specificity of enzymes achieved?   (2 marks)

(c) How does the induced-fit theory of enzyme action differ from the lock-and-key theory? (2 marks)

℮ This question is primarily testing recall. However, many students may not score full marks as they will not give answers that are detailed enough.

Student A

(a) They speed up chemical reactions.
(b) Their active sites.
(c) The active site changes.

℮ **2/6 marks awarded** (a) Student A only scores 1 mark due to not including enough detail. He or she has not looked at the mark allocation for the question — a 2-mark question requires at least two points. The student should have included the fact that enzymes speed up reactions by lowering the activation energy or that enzymes are unchanged at the end of the reaction. (b) Again, Student A only scores 1 mark by not developing his or her answer. (c) Student A scores no marks here due to a lack of specificity and no comparison. The question is asking for a difference and this must be given in the answer.

Student B

(a) They speed up reactions in living organisms by lowering the activation energy of the reaction and the enzymes are unchanged at the end of the reaction.
(b) The active site makes it specific.
(c) In the lock-and-key theory the active site is a complementary shape to the substrate. In the induced-fit theory the active site changes shape to fit the substrate.

ⓔ **5/6 marks awarded awarded** (a) This is a very good detailed answer that gains 2 marks. (b) This answer only contains one point so cannot score 2 marks. Active site gains 1 mark while mention of its specific tertiary structure would have gained the second mark. (c) Student B has correctly explained the difference between the two theories and so gains both marks.

Question 11

The graph below shows the mass of product formed by an enzyme-controlled reaction over time.

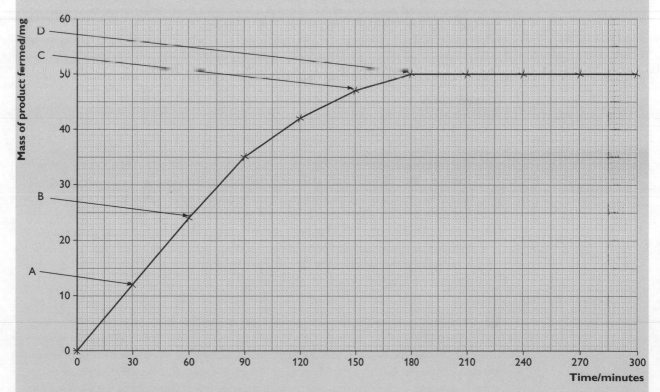

(a) Calculate the rate of reaction between:
  A and B $mg\,min^{-1}$
  C and D $mg\,min^{-1}$ (2 marks)

(b) Explain why these two values are different.   (3 marks)

(c) Explain the shape of the curve after point D.   (2 marks)

ⓔ Questions asking you to calculate rate are quite common in BY1 exams. It is an easy calculation — the rate is found by dividing the mass of product by the time taken to produce it. The units will be whatever the product or substrate is measured in over the unit of time. In this case it is given for you, but if it is not you should always include it. The rest of the question is fairly straightforward. The main issue with a product over time graph like this one is that students often mix this graph up with the rate of reaction/substrate concentration graph.

Student A

(a) A–B $12/30 = 0.4\,\mathrm{mg\,min^{-1}}$
C–D $3/30 = 0.1\,\mathrm{mg\,min^{-1}}$

(b) The rate of reaction was faster at A–B than C–D. This is because C–D is closer to the maximum rate of reaction.

(c) After point D all the active sites are full all the time. This means the reaction has reached its maximum rate.

ⓔ **2/7 marks awarded** Student A's answer reflects what can happen if you fail to identify the type of graph you are being asked to look at. This student clearly learned the shapes of the different rate of reaction graphs and the facts that accompany them (the graph levels out due to all the active sites being full all the time) and then has just reeled if off here. This single mistake has led to Student A only scoring 2 marks and underlines the necessity of not just learning the specification content off by heart but really thinking about it and trying to understand it. The only marks this student scored were for the correct rate calculations in part (a).

Student B

(a) A–B $12/30 = 0.4\,\mathrm{mg\,min^{-1}}$
C–D $3/30 = 0.1\,\mathrm{mg\,min^{-1}}$

(b) The rate of reaction was fastest at A–B as at this point there was a large concentration of substrate so the enzyme concentration was the limiting factor. At C–D the rate of reaction is slower. This is because the concentration of substrate has decreased as it has been converted into products. This means there is a reduced chance of a collision between an active site and a substrate molecule therefore less product is produced per minute.

(c) After point D the graph is flat. This means the rate of reaction is 0 as all the substrate has been converted into product.

ⓔ **7/7 marks awarded** Student B's answer illustrates the difference that good understanding and preparation for exams can make. This student has obviously thought carefully about what the graph showed and was able to score all the marks available for this question. (a) Both rates are calculated correctly so this gets 2 marks. (b) This is an excellent answer that scores all 3 marks available. There was 1 mark for explaining that A–B was faster due to there being a large concentration of substrate at this point, near the start of the reaction. Student B scored the other 2 marks by identifying that the C–D rate of reaction is slower because some of the substrate has been converted to products so there is less chance of a successful collision between a substrate molecule and an active site. (c) Student B scores both marks by saying the rate of reaction is zero because all the substrates have been converted into products.

Question 12

The graph below shows the rate of a reaction catalysed by the enzyme lactase at different pHs, when it is immobilised (○) and when it is free in solution (◉).

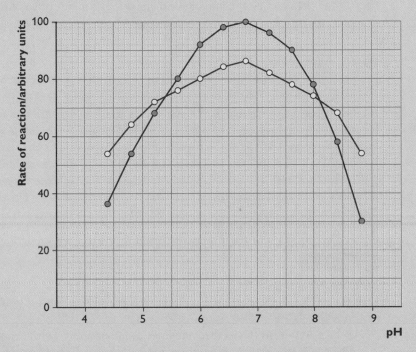

(a) Lactase hydrolyses the glycosidic bond in lactose. What will be the products of this reaction?

(2 marks)

(b) Explain the rate of the free lactase reaction at:
   pH 6.8
   pH 4.4

(3 marks)

(c) Using information from the graph, describe two effects of immobilisation on the rate of reaction.

(2 marks)

ⓔ It is important in questions involving graphs to study them carefully. You are looking for what the graph is showing you (this will be the answer to 'describe' questions) and the underlying theory that explains the shape of the curve (this will answer an 'explain' question). Part (c) asks you to use information from the graph — make sure you do this and do not bring in other additional information that might not be relevant.

(a) Glucose and fructose.
(b) pH 6.8 This is the enzyme's optimum pH so at this point the rate of reaction is the fastest.
pH 4.4 The rate of reaction is low as the enzymes are becoming denatured.
(c) At the optimum pH the immobilised enzyme has a lower rate of reaction than the non-immobilised enzyme. The immobilised enzymes survive better at extreme pHs.

e **4/7 marks awarded** (a) Student A only scores 1 mark as he or she has failed to identify galactose as the monosaccharide that bonds with glucose to form the disaccharide sucrose. (b) For the first point Student A has given a good answer and scores the 1 mark available. However, for the second part of the question the student scores only 1 of the 2 marks available. Student A has used the term denatured, which scores 1 mark, but has failed to explain what it means. In a question like this you should always explain terms like denatured. (c) The first part of this answer is correct, but the second part lacks detail and uses the term 'survives' so Student A only gets 1 of the 2 marks available. The use of survive illustrates a common mistake that students make when answering enzyme questions. Words such as 'live', 'kill' and 'survive' are often used in answers. These imply that an enzyme' is alive and should not be used in exam question answers — enzymes are not living, they are just chemicals.

(a) Glucose and galactose.
(b) pH 6.8 This is the enzyme's optimum temperature so at this point the rate of reaction is the fastest.
pH 4.4 At this pH the lactase enzymes are denatured. This pH has caused a change in the tertiary structure of the active site so the substrates can no longer bind so the rate of reaction has decreased.
(c) The maximum rate of reaction of the immobilised lactase is lower than the free lactase. However, at a pH above 8.1 the immobilised lactase has a higher rate of reaction than the free lactase.

e **7/7 marks awarded** (a) These are the two correct monosaccharides that make up lactose so score 2 marks. (b) This is an excellent answer to this question — the pH 4.4 answer is very detailed but it is better to write too full an answer than too little and it is always worth explaining what denaturing means. (c) Both of these differences are correct and this answer scores 2 marks. It would also have been possible to score marks for stating that the rate of reaction of the immobilised enzyme is higher at pHs below 5.4.

Question 13

Enzymes are affected by two different types of inhibitors — competitive inhibitors and non-competitive inhibitors.

(a) Give an example of a competitive inhibitor and an example of a non-competitive inhibitor. (2 marks)

(b) Enzyme inhibitors work by preventing the formation of enzyme–substrate complexes. Explain the different ways competitive and non-competitive inhibitors achieve this. (4 marks)

(c) The graph below shows the effect of increasing substrate concentration of a reaction when the inhibitors a or b are present in the solution. Describe and explain the differing effects of a and b on the rate of reaction. (3 marks)

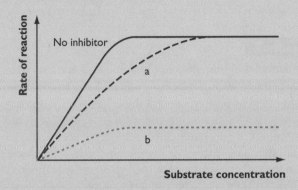

e This question begins with some straightforward recall. You should aim to score full marks on questions such as part (a) because they require no further thought or analysis — you just need to remember the facts. Students often struggle to recall an example of a competitive inhibitor in particular. Part (b) should also be relatively straightforward — just make sure you explain both processes and use all appropriate key words. Part (c) is a challenging question as students often find this graph difficult to interpret. This is a describe and explain question so even if you are struggling to fully explain what is happening you can still pick up marks by describing what the graph is showing.

Student A

(a) An example of a competitive inhibitor is malonic acid and an example of a non-competitive inhibitor is cyanide.

(b) A is a competitive inhibitor. It competes with the substrate for the active site as it is the same shape as the substrate. A non-competitive inhibitor causes a change in the shape of the active site of the enzyme by binding to a site other than the active site.

(c) As substrate concentration increases the reaction with the competitive inhibitor does eventually reach its maximum rate. However, the reaction with the non-competitive inhibitor never reaches its theoretical maximum rate of reaction no matter how much substrate is added.

ⓔ 5/9 marks awarded (a) These are two correct examples so gain 2 marks. (b) Student A scores no marks for this explanation of competitive inhibition. Saying the inhibitor 'competes' with the substrate is not enough to score a mark — this is really just restating information that is in the start of the question (a 'competitive inhibitor' would obviously 'compete'). The student has also made a common mistake in describing the inhibitor as being the same shape as the substrate. The inhibitor is not the same shape as the substrate, it is a similar shape. The next part of this answer does score 2 marks for describing the non-competitive inhibitor binding to a site other than the active site and changing its shape. It would have been better if the student had gone on to say this prevents the formation of enzyme–substrate complexes as this is what the question was initially asking for. You should always aim for the most complete and full answer possible to ensure that you pick up all the marks available. (c) While this answer only gains 1 mark of the 3 available it is a good example of how to score marks in a difficult question even if you do not fully understand it. This student cannot answer the explain part of the question, but rather than just missing the whole question out or writing a rambling answer that he or she knows is not right Student A has focused on the part of the question he or she can do, writing a good answer and getting a mark.

Student B

(a) Examples of inhibitors are malonic acid and cyanide.
(b) A competitive inhibitor is a similar shape to the substrate of the reaction. The competitive inhibitor enters the active site of the enzyme. This prevents the substrate from entering, preventing the formation of an enzyme–substrate complex. A non-competitive inhibitor does not bind to the active site but to another part of the enzyme known as the allosteric site. This causes a change in the tertiary structure of the active site that means the substrate can no longer bind. This therefore prevents the formation of an enzyme–substrate complex.
(c) As the substrate concentration increases the effect of the competitive inhibitor is reduced and eventually the maximum rate of reaction is reached. This is because as the concentration of substrate increases there is a much greater chance of a substrate/active site collision as opposed to an inhibitor/active site collision. The non-competitive inhibitor reduces the number of available active sites. This means that the rate of reaction can never reach its maximum.

ⓔ 7/9 marks awarded (a) Student B has made a basic but common error in this part of the question. In his or her rush to write down the correct answer the student has ended up scoring no marks out of the 2 available. While both these answers are correct, Student B has not identified which is a competitive inhibitor and which is a non-competitive inhibitor and so the examiner cannot award any marks. This underlines the importance of ensuring that you think carefully about your answers and read them back to make sure they are as complete as possible. (b) This is an excellent, detailed answer that scores all 4 marks available, 2 marks for the description of the competitive inhibitor and 2 marks for the non-competitive inhibitor. (c) This is a good answer that scores all 3 marks available — 1 mark for the description of the differing effects of the inhibitors and 2 marks for explaining the effects.

Nucleic acids

Question 14

(a) Why can DNA be described as a polymer? (2 marks)

(b) Explain three ways in which mRNA differs from DNA. (3 marks)

(c) The nitrogenous bases in a sample of DNA were analysed. Thymine was found to make up 13% of the nitrogenous bases. Calculate the percentage of cytosine in the sample. Explain how you arrived at your answer. (3 marks)

ⓔ This question requires mainly factual recall and a fairly simple calculation so should not prove too difficult. The DNA questions on the BY1 exam tend to be the source of fairly easy marks.

Student A

(a) It is made up of monomers.
(b) DNA has thymine nitrogenous base, RNA has uracil.
DNA nucleotides have deoxyribose, RNA nucleotides have ribose.
RNA is single stranded, DNA is double stranded.
(c) 13% of thymine means there must be 13% cytosine.

ⓔ **4/8 marks awarded** (a) This answer is correct but a single point like this will only score 1 mark. (b) This is an excellent answer — it makes three good comparisons so scores all 3 marks available. (c) Student A has mixed up the complementary base pairs and believes thymine pairs with cytosine. This means this answer scores no marks. The student could have picked up this error when checking through the exam paper by looking at the mark allocation for this question. If the answer to this question were so simple it would not be worth 3 marks. Student A has also not explained how he or she arrived at this answer. The term complementary base pairing was needed to get this mark.

Student B

(a) It is made up of many monomers known as nucleotides.
(b) RNA contains the nitrogenous base uracil while DNA contains the nitrogenous base thymine. DNA has a different pentose sugar to RNA and it is double stranded.
(c) There is complementary base pairing between adenine and thymine and cytosine and guanine.
So amount of thymine = amount of adenine = 13+13 = 26%
Therefore 100 – 26% = 74% amount of cytosine and guanine
As amount of cytosine = amount of guanine, % of cytosine = 74/2 = 37%

ⓔ **6/8 marks awarded** (a) This is a detailed answer that scores 2 marks. (b) The first point is a good comparison and scores 1 mark. The next two points are both correct, but as they are not comparative statements they score no marks. (c) This excellent answer scores all 3 marks. First,

Student B has stated the fundamental concept behind answering a question like this. As there is complementary base pairing the proportions of the rest of the bases can be calculated from the amount of one base, as is done in this student's answer.

Cell division

Question 15

(a) In which stages of mitosis do (i)–(iv) occur?

   (i) **The nuclear membrane reforms.**   (1 mark)

   (ii) **Spindle fibres contract and pull sister chromatids to the opposite poles of the cell.**   (1 mark)

   (iii) The nuclear membrane breaks down.   (1 mark)

   (iv) Spindle fibres join to the chromosomes at the centromere.   (1 mark)

(b) **A student was observing some plant tissue in which cells were undergoing mitosis. She saw a large number of cells that looked like A, but very few that looked like B. Identify stages A and B and explain the difference in numbers she saw.**   (4 marks)

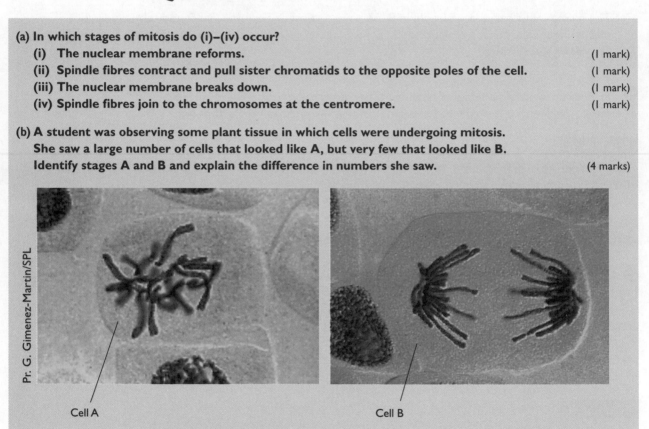

Pr. G. Gimenez-Martin/SPL

Cell A   Cell B

(e) Mitosis questions often ask you to match up what is happening in a phase with its correct name. It is therefore important to have a firm grasp of exactly what occurs in each stage. This question also involves looking at photographs of real plant cells undergoing mitosis. Students often find this challenging because they do not always look like the diagrams we usually look at. In a cell division photograph the key things are the chromosomes (the black squiggles) — look at how they are behaving and relate that to what you know the chromosomes do in each of the stages of mitosis.

Student A

(a) (i) Telophase
   (ii) Metaphase
   (iii) Prophase
   (iv) Anaphase
(b) Cell A is in prophase and cell B is in metaphase. Prophase is a much longer phase than metaphase so you would expect to see more cells in prophase in a sample.

📋 **4/8 marks awarded** (a)(i)–(iv) Student A only scores 2 of the 4 marks available (for (i) and (iii)) as he or she has mixed up anaphase (the correct answer for (ii)) and metaphase (the correct answer for (iv)). It is important to have what occurs in each stage clear in your mind. (b) Mixing up metaphase and anaphase again loses the student marks here. The chromosomes separate and are pulled to the opposite poles of the cell during anaphase. Student A picks up 2 marks for correctly stating that A is a cell in prophase and that prophase is a long phase so you would expect to see a large number of cells in prophase.

Student B

(a) (i) Telophase
   (ii) Anaphase
   (iii) Prophase
   (iv) Metaphase
(b) Cell A is in prophase while cell B is in anaphase. Prophase is a very long phase therefore more of the cells at this point in time are in prophase.

📋 **7/8 marks awarded** (a)(i)–(iv) All four of these stages are correct so Student B scores all 4 available marks. (b) Student B has correctly identified both diagrams and has stated that prophase is a long phase therefore more of the cells would be in prophase at any one time. However, Student B does not score the fourth mark available for this question as he or she has not made a comparison with cell B to say anaphase is a much quicker phase so you would only expect to see a small number of cells in anaphase at any one time.

Question 16

The graph below shows the change in the amount of DNA during the cell cycle.

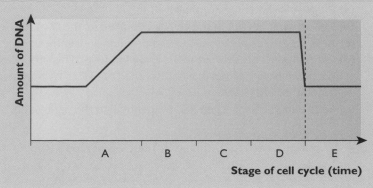

(a) Identify stage A. (1 mark)

(b) Explain the change in DNA during stage A. (1 mark)

(c) Name two other processes that occur during stage A. (2 marks)

(d) Explain the change in DNA during stage D. (2 marks)

(e) How does mitosis ensure that the daughter cells produced are identical to the parent cell? (1 mark)

🅔 The questions involving this graph are quite challenging. As mentioned before, it is vital to ensure you work out what a graph is showing before starting your answer, as a single misunderstanding can result in you losing a large number of marks.

Student A

(a) Stage A is interphase.
(b) DNA replication has occurred so the amount of DNA has doubled.
(c) ATP is produced and protein synthesis occurs.
(d) The amount of DNA halves as each daughter cell receives one copy.
(e) Mitosis ensures that the daughter cells are exactly the same in every way to the parent cell.

🅔 **5/7 marks awarded** (a) and (b) Student A has correctly identified interphase and the DNA doubling due to DNA replication so gains both marks. A common mistake on this part of the question is to identify stage A as a stage in mitosis — this would lead to losing the mark in (a) and the 2 marks for (c). The key point to remember here is that DNA replicates during interphase before mitosis begins, therefore stage A must be interphase. (c) Both of these processes occur in interphase so this answer gets both marks. (d) This answer can only get 1 mark as it only makes one point (that each daughter cell receives a copy of the DNA). The second mark would be awarded for stating that cytokinesis occurs splitting the cell into two daughter cells. Try to use keywords like cytokinesis whenever possible, because there are often marks available for their correct use. (e) This is another example of an answer that while correct does not contain enough detail to be awarded a mark. The key point here is that mitosis produces daughter cells that are genetically identical to the parent cell. This is important because it ensures genetic stability.

Student B

(a) Stage A is showing interphase.
(b) The amount of DNA in the cell has doubled as the DNA has replicated.
(c) New organelles are produced and proteins are synthesised.
(d) Cytokinesis has occurred at the end of mitosis. This produces two daughter cells, each of which receives one copy of the DNA, This is shown on the graph by the amount of DNA halving.
(e) Each daughter cell is genetically identical to the parent cell.

🅔 **7/7 marks awarded** (a), (b) and (c) These are all good answers and Student B gets all 4 marks available. (d) This is a detailed, well-developed answer that gains both marks. (e) Using the key term genetically identical ensures the student gets the mark for this question.

Question 17

Fill in the blanks in the paragraph below on the differences between mitosis and meiosis.

Mitosis produces daughter cells that are genetically _____(a)_____ to each other and to the parent cell. This is important as it ensures ____(b)_____ stability. Mitosis is used for ____(c)_____, replacement of dead cells and ____(d)_____ of damaged tissues in living organisms. Meiosis produces four ____(e)_____ daughter cells which have half the number of chromosomes found in a ____(f)_____ cell. Genetic variation occurs in meiosis due to _____(g) _____ (swapping of genes between homologous chromosomes) and _____(h) _____ which occurs at metaphase I.

(8 marks)

ⓔ Fill in the blanks questions can be a source of fairly easy marks. However, one mistake can lead to further incorrect answers, which can soon add up. The key to this type of question is not to just jump straight in and start writing down answers quickly. Read the paragraph through carefully at least twice before writing any answers down. Also, be wary of basing an answer on a previous answer (as (a) is...therefore (b) must be...). If your first answer is wrong your reasoning to get the second answer will also be wrong and you will have lost 2 marks.

Student A

(a) identical
(b) genetic
(c) repair
(d) growth
(e) different
(f) mitotic
(g) crossing over
(h) independent assortment

ⓔ **4/8 marks awarded** The first two answers are correct so the student scores 2 marks. Student A has then got (c) and (d) (repair and growth) the wrong way round. This is an example of answering a question too quickly and not fully thinking it through — (c) could have been repair but the part of the question after (d) 'of damaged tissues' clearly points to (d) being repair. Student A should have gone back and thought about what (c) could have been instead. (e) and (f) are also incorrect. The clue to these answers is the statement 'daughter cells which have half the number of chromosomes' — this is pointing you towards haploid and diploid (the two correct answers). Examiners are not going to give you answers to fill in the blanks questions. Mitotic sounds like a good answer, but when the sentence starts with 'Meiosis produces...' it is unlikely that an answer is going to be the name of the other type of cell division. (g) and (h) are both correct.

Student B

(a) identical
(b) genetic
(c) growth
(d) repair
(e) haploid
(f) diploid
(g) independent assortment
(h) crossing over

ⓔ **6/8 marks awarded** Answers (a) to (f) are all correct. However, Student B has mixed up independent assortment and crossing over and so loses 2 marks. When learning the sources of variation in meiosis students often just remember the terms, not what they mean. When a question like this comes up they then put down the two terms they have learned but without knowing what they mean. This is an example of how a simple mistake like mixing up two terms can cost you 2 marks. The best way to avoid this is thorough revision and reading back carefully over your answers.

Essay questions

Question 18

Describe the structure of lipids and their importance to living organisms. (10 marks)

ⓔ The essay question is a good opportunity to score marks. You will be given a choice of two questions so you will be able to pick the one you feel most comfortable with. The key to writing a good BY1 essay is to be detailed and concise. It is often possible to hit three of the mark points and get 3 marks with a single sentence. Do not waffle and stick to the subject of the question.

Student A

Lipids are made up of the elements carbon, oxygen and hydrogen. Lipids are polymers. Polymers are large molecules that are made up of many monomers. The monomers in lipids are joined to each other by ester bonds. The monomers in a triglyceride are three glycerol molecules and a fatty acid. Fatty acids can be either saturated or unsaturated. This depends on double bonds within the fatty acid. A saturated fatty acid has no double bonds while an unsaturated fatty acid does have double bonds. There are also phospholipids. Phospholipids are made up of two fatty acids joined to a glycerol phosphate head. The tails of the fatty acid are hydrophobic and the heads are hydrophilic — this is important in the formation of cell membranes. Lipids are also used for energy storage in animals.

ℯ 4/10 marks awarded The description of the elements of a lipid gains 1 mark, but Student A goes on to make the incorrect point that lipids are polymers. He or she then mixes the incorrect monomers and polymers ideas with correct statements of ester bond name and the components of a triglyceride. Because Student A has mixed incorrect and correct statements he or she does not gain any marks for stating the name of the bond in a lipid is an ester bond and for describing the components of a triglyceride. Student A gets 1 mark for saying fatty acids are saturated or unsaturated, but does not explain the difference properly so does not get a second mark. A saturated fatty acid has no carbon to carbon double bonds, though it does have a double bond between an oxygen and a carbon. Student A correctly describes the structure of a phospholipid so gains 1 mark, but has mixed up the properties of the hydrophilic heads and the hydrophobic tails so does not get a mark for this. Student A then gives a correct function for a lipid and this scores 1 mark. This question was asking about structure and function so Student A should have tried to include a few more functions.

Student B

Lipids are made up of two subunits: glycerol and fatty acids. Three fatty acids join to one glycerol molecule to form a triglyceride. Three condensation reactions occur to form three ester bonds which join the glycerol to the fatty acids. Three molecules of water are also formed. Fatty acids can be classified as saturated or unsaturated. A saturated fatty acid has no carbon to carbon double bond within its hydrocarbon chain — this therefore means it has the maximum amount of hydrogen atoms. An unsaturated fatty acid does have carbon to carbon double bonds within its hydrocarbon chain, therefore it does not have the maximum number of hydrogen atoms. Phospholipids are a special type of lipid which only have two fatty acids bonded to the glycerol molecule. They also have a phosphate group attached to the glycerol. This makes the head of the phospholipid hydrophobic and the fatty acid tails hydrophilic. Lipids are very important in living organisms. They are used in energy storage — they store twice as much energy as the same mass of carbohydrate. This makes them a more efficient store of energy. Lipids can also be used as signalling molecules, for example the steroid hormone dopamine.

ℯ 9/10 marks awarded This is an excellent answer. Student B gets 3 marks for describing the structure of a triglyceride, its formation and the name of the bond. Stating that fatty acids can be saturated or unsaturated and explaining the structural difference between the two scores 2 marks. The explanation of the structure of the phospholipid and its hydrophilic and hydrophobic nature gets another 2 marks. Student B then gets a further 2 marks for describing a function of lipids and explaining it in more detail (energy storage and why lipids are more efficient energy stores than carbohydrates). However, the student did not score a mark for saying that lipids can be used as signalling molecules. This information is not on the BY1 specification. Sometimes the examiner will award 1 mark for additional information not on the specification, but it is best to stick to the specification. In this case, the student has incorrectly identified dopamine as a steroid hormone, so does not score a mark.

Question 19

Explain the importance of water to living organisms. (10 marks)

ⓔ Water is an area of the specification that students often do not spend much time on, but it does come up in the BY1 exams and if you have not revised it properly you could lose a lot of marks. When writing an essay on a topic like water it is important to be specific and stick to the specification content, for example, writing large amounts about the physiological effects of dehydration, which is not on the specification, will only get one 'any other valid point' mark at most.

Student A

Water is an important substance for living things. It is very important that organisms drink enough water. Water is a solvent; this means it dissolves lots of different things. This makes water important in the transport of molecules such as glucose dissolved in blood. Water also has a very high latent heat and a very high specific heat. This is important as the high latent heat means that it takes a large amount of energy to raise the temperature of water by one degree Celsius. This provides a relatively constant internal environment for cells. The high specific heat means a large amount of energy is used to evaporate and evaporation of sweat cools humans down.

A water molecule is made up of an oxygen atom and two hydrogen atoms. Water is a polar molecule as the oxygen atoms have a slight negative charge and the hydrogen atoms have a slight positive charge. Water molecules can form hydrogen bonds between a hydrogen atom and an oxygen atom from two different water molecules. These hydrogen bonds stick the water molecules together and give high surface tension. Water is also transparent which allows sharks to hunt their prey.

ⓔ **6/10 marks awarded** This answer is characterised by a lack of detail and an easily made mistake which prevented the student scoring a good 8/10 mark as opposed to a relatively poor 6/10 mark. The comment on drinking enough water is not an A-level standard answer and does not score a mark, nor does describing water as dissolving a lot of things — the term universal solvent was needed here. Student A does score 1 mark for an example of the importance of water being a solvent by describing glucose being transported in the blood. The student also gets 2 marks for stating that water has both a high latent heat and a high specific heat. However, Student A then mixes up the meanings of specific heat and latent heat. This one mistake costs him or her 2 marks because, apart from the mix up, the student's descriptions of water's high specific heat providing a constant internal environment for cells and the cooling effect caused by the high latent heat are good. Student A scores 3 marks for identifying water as a polar molecule, having an uneven distribution of charge and the formation of hydrogen bonds. However, the student fails to score any marks in the rest of the answer. Saying that hydrogen bonds stick the water molecules together is not enough to get a mark — the keyword cohesion was needed. Student A is correct in stating that water has a high surface tension, but to score a mark for this fact the student needed to link it to its importance to living organisms, for example that it allows some insects to walk across the surface of water. While it is true that the transparent nature of water allows sharks to hunt it is not the most important benefit of water's transparency. To score the mark here the student should have stated that the transparency of water allows light to reach aquatic plants to allow them to photosynthesise.

Student B

Water is of vital importance to all life. It is known as the universal solvent as it dissolves such a wide range of different solutes. This is especially important as it means chemical reactions (such as the enzyme-catalysed ones that occur in living cells) can occur in solution. It also allows water to act as a transport medium for complex multicellular animals, for example transport of photosynthetic products in the phloem of terrestrial plants. Water is a polar molecule as it has an uneven distribution of charge (the hydrogen has a slight positive charge while the oxygen atom has a slight negative charge). This uneven distribution of charge means water forms hydrogen bonds. This leads to cohesion between water molecules. This allows large water columns to be pulled up the xylem of plants. Water has a high specific heat — this means that a large amount of energy is required to raise the temperature of water. This means bodies of water provide a relatively stable environment for aquatic animals. When water freezes to form solid ice it is less dense than liquid water so floats on top of the water. Again this provides an insulating layer, which allows aquatic organisms to survive beneath the ice.

🅮 **10/10 marks awarded** This is an excellent answer that would actually score more than the maximum 10 marks that can be awarded. Student B gets marks for stating water is a universal solvent, explaining this is important to allow chemical reactions to occur in solution and the importance of it in transport in plants. The student gains 4 marks for the description of the polar nature of water, the uneven distribution of charge, the formation of hydrogen bonds and the importance of cohesion. This is a good example of how a few detailed, concise sentences can give you a large number of marks in an essay. Student B's points on the high specific heat of water score another 2 marks. The last two sentences are a good example of describing a feature of water and clearly explaining its importance to living organisms, scoring another 2 marks.

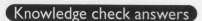

Knowledge check answers

1 It has an uneven distribution of charge — the oxygen atom is slightly negative and the hydrogen atom is slightly positive.
2 Hydrogen bonds form between the oxygen atom of one water molecule and the hydrogen atom of another.
3 It dissolves a large number of different solutes.
4 High specific heat capacity
5 When the water evaporates it provides a cooling effect, e.g. in sweating.
6 If a plant does not have enough magnesium it cannot make the green pigment chlorophyll.
7 Two glucose molecules and one molecule of water.
8 If they were soluble, structures made up of polysaccharides would dissolve in water.
9 Glycogen
10 Beta glucose
11 Hydrogen bonds
12 One OH group on its beta glucose monomers is replaced with an amino acid.
13 Unsaturated fatty acids have carbon to carbon double bonds; saturated fatty acids do not.
14 Ester bond
15 20
16 The type, number and sequence of amino acids

17 Alpha helix and beta pleated sheet
18 Haemoglobin
19 To produce ATP.
20 To provide a large surface area for ATP synthesis.
21 Because it has ribosomes along its length.
22 Chlorophyll
23 Intrinsic proteins lie across both layers of the phospholipid bilayer. Extrinsic proteins are in one of the layers or on the surface of the membrane.
24 Facilitated diffusion requires a channel or carrier protein while diffusion does not.
25 It becomes plasmolysed.
26 They lower the activation energy of the reaction.
27 Globular proteins
28 Pepsin is found in the stomach where the pH is low.
29 All the available enzyme active sites are full all the time.
30 A competitive inhibitor binds to the active site, preventing the substrate from entering. A non-competitive inhibitor binds to an area of an enzyme away from the active site; this changes the shape of the active site so the substrate can no longer bind.
31 Adenine and guanine
32 Pentose sugar (deoxyribose) and phosphate
33 In the nucleus.
34 A centromere
35 27